Tamer Ezzat Youssef

Advanced Designed Phthalocyanine Materials for Nonlinear Optics

Tamer Ezzat Youssef

Advanced Designed Phthalocyanine Materials for Nonlinear Optics

Novel Designed Phthalocyanines and their possibilities to use as Optical Limiters

Südwestdeutscher Verlag für Hochschulschriften

Imprint
Any brand names and product names mentioned in this book are subject to trademark, brand or patent protection and are trademarks or registered trademarks of their respective holders. The use of brand names, product names, common names, trade names, product descriptions etc. even without a particular marking in this work is in no way to be construed to mean that such names may be regarded as unrestricted in respect of trademark and brand protection legislation and could thus be used by anyone.

Publisher:
Südwestdeutscher Verlag für Hochschulschriften
is a trademark of
Dodo Books Indian Ocean Ltd., member of the OmniScriptum S.R.L Publishing group
str. A.Russo 15, of. 61, Chisinau-2068, Republic of Moldova Europe
Printed at: see last page
ISBN: 978-3-8381-2378-3

Zugl. / Approved by: Tübingen: Faculty of Chemistry and Pharmacy, Institute of Organic Chemistry, Eberhard-Karls -Tübingen University, Germany- Diss., 2004

Copyright © Tamer Ezzat Youssef
Copyright © 2011 Dodo Books Indian Ocean Ltd., member of the OmniScriptum S.R.L Publishing group

Preface

Nowadays, phthalocyanines (Pcs) form an important group of aromatic macrocycles, dependant on a delocalized 18-π electron system that belongs to the most modern functional materials in scientific research. They play an important role in many advanced applications and modern technologies, mostly by virtue of their characteristic optical absorption and high chemical stability. In addition, they can find commercial applications as photoconductors in xerographic machines, electrochromic displays, photovoltaic materials in solar cells, systems for fabrication of light emitting diodes (LED), optical limiters, dyes for recording layers in recordable digital versatile discs (DVDs), liquid crystalline materials, organic conductors, sensitizers for photodynamic cancer therapy (PDT) and diverse catalytic systems.

I invited world-recognized scientists who are specialists in design and synthesis of phthalocyanine materials field to contribute specific chapters to this monograph. This was in order to cover major aspects of non-linear optics from the fundamentals to various applications.

The monograph in general presents a balanced and informative description of progress in molecular non-linear optics till 2004, includes the preparation of stable phthalocyanine material systems like actual nonlinear materials, building of a knowledge base and an intuitive understanding of structure/function relationships in simple molecular systems.

This monograph contains six chapters with 21 Schemes, 39 Figures and 7 Tables. The first chapter provides an extensive introduction to the discovery of phthalocyanines, the various general synthetic routes and types of substituted phthalocyanines. Surveys the current advances of the nonlinear optical (NLO) properties of phthalocyanines. It examines the methods employed to compute the properties and introduces readers to the theory of linear and nonlinear optical phenomena: mechanism and model. Chapter 2 devoted to the aim of work, it also point out some of the present challenges for the accurate calculation of the NLO properties of interest. Chapter 3 divided into two parts, Part 1 is concerned with the design, synthesis and characterization of fully conjugated soluble binuclear and trinuclear phthalocyanines.

Part 2: Covers the recent NLO properties of target phthalocyanines. Strategies to develop novel NLO materials are also discussed along with the Z-scan technique. Chapter 4 comments on the course of reactions for these compounds starts with the synthesis of unsymmetrically substituted phthalocyanine precursors to yield the target compounds. Chapter 5 provides an details of the methods of synthesis used. Finally, Chapter 6 is literature. It is hoped that the topics provided in this work will be valuable as tutorials for the beginning researchers and as a desktop reference for the advanced researchers for some time to come. Your comments are very important to me, so please feel free to e-mail your suggestions to: Tamer.Ezzat@Uni-tuebingen.de.

Dr.rer.nat.Tamer E. Youssef

Hannover, January, 2011

Contents

Chapter 1

1. Introduction 1

2. Phthalocyanines 2
   2.1. General Remarks 2
   2.2. Discovery of phthalocyanines (Pc`s) 2
   2.3. Structures of phthalocyanines 2
   2.4. Syntheses of phthalocyanine 3
      2.4.1. General synthetic routes 3
      2.4.2. Electrochemical synthesis of phthalocyanines 5
   2.5. Soluble phthalocyanines 5
      2.5.1. Tetrasubstituted phthalocyanines 6
      2.5.2. Octasubstituted phthalocyanines 7
      2.5.3. Unsymmetrically substituted phthalocyanines 8

3. Ladder polymers 11
   3.1. Synthesis of Ladder polymers 11
   3.2. Selected dimers, trimers and oligomers of phthalocyanines 12

4. Stacked phthalocyanines 13
   4.1. Nonbridged stacked polyphthalocyanines 13
   4.2. Bridged stacked polyphthalocyanines 14

5. Optical limiting: A nonlinear optical effect 15
   5.1. Optical limiting phenomena 16
   5.2. Nonlinearity of the optical effect 16
   5.3. Optical limiting: mechanism and model 17
   5.4. Phthalocyanines for optical limiting 20

Chapter 2

Aim of Work 24

Chapter 3

1. Synthesis of fully conjugated soluble binuclear phthalocyanine 27

    1.1. Synthesis of dienophilic phthalocyanines 28

    1.2. Synthesis of asymmetrically substituted mono dienophilic
    Nickel(II)phthalocyanine with AAAB-Symmetry: 30

    1.2.1 Spectroscopic characterization... 30

    1.3. Synthesis and spectroscopic characterization of Pc-tetracyclone adduct 31

    1.4. Synthesis of planar binuclear metallophthalocyanine dimer 33

        1.4.1. Spectroscopic characterization.. 35

    1.5. Synthesis of dehydrated planar metallophthalocyanine dimer 37

        1.5.1. Spectroscopic characterization.. 37

2. Synthesis of planar trinuclear metal free Phthalocyanine 39

    2.1. Synthesis of symmetrically substituted bis-dienophillic
    metal free phthalocyanine with ABAB-symmetry:... 39

    2.1.1. Spectroscopic characterization of phthalocyanine 40

    2.2. Synthesis and spectroscopic characterization of Pc-tetracyclone bisadduct 40

    2.3. Synthesis of planar trinuclear metal free-phthalocyanine trimer 41

    2.3.1 Spectroscopic characterization of metallophthalocyanine trimer 43

3. Synthesis of Pyridine *N*-oxide adducts 45

    3.1. Spectroscopic characterization of pyridine *N*-oxide adducts............................. 47

**4. Synthesis and properties of Phthalocyaninatoindium(III)acetylacetonates
designed for optical limiting purposes** 50

    4.1. Synthesis of Octa-(2-ethylhexyloxy)phthalocyaninatoindium(III)acetylacetonate 51

        4.1.1. Spectroscopic characterization.. 51

    4.2. Synthesis of Tetra-*tert*-butylphthalocyaninatoindium(III)acetylacetonate 52

        4.2.1. Spectroscopic characterization 53

    4.3. Synthesis of trinuclear metallophthalocyanine trimer 54

        4.3.1. Spectroscopic characterization 55

5. Nonlinear optical properties of Phthalocyaninatoindium(III)acetylacetonates.................. 57

    5.1. Z-Scan measurements 58

    5.2. Optical limiting measurements 61

6. Synthesis of binuclear Ruthenium Phthalocyanine	64
6.1. Synthesis of Hexadecafluoro(phthalocyaninato)ruthenium(II)	66
6.2. EXAFS spectroscopic measurements	66
6.3. Spectroscopic characterization	69
6.4. Interaction with oxygen in solution	69
6.5. Magnetic measurements	70

Chapter 4

Summary	72

Chapter 5

Experimental Part	74
1. General Comments	74
2. Synthesis	77
2.1. Synthesis of precursors	77
2.1.1. 1,2-Dicyano-4,5-bis(2-ethylhexyloxy)benzene	77
2.1.1.1. 1,2-Bis(2-ethylhexyloxy)benzene	77
2.1.1.2. 1,2-Dibromo-4,5-bis(2-ethylhexyloxy)benzene	77
2.1.1.3. 1,2-Dicyano-4,5-bis(2-ethylhexyloxy)benzene	78
2.1.2. 6,7-Dicyano-1,4-epoxy-1,4-dihydronaphthalene and 6,7-Dicyano-1,4-epoxy-1,4-dimethyl-1,4-dihydronaphthalene	79
2.1.2.1. 6,7-Dibromo-1,4-epoxy-1,4-dihydronaphthalene and 6,7-Dibromo-1,4-epoxy-1,4-dimethyl-1,4-dihydronaphthalene	79
2.1.2.2. 6,7-Dicyano-1,4-epoxy-1,4-dihydronaphthalene and 6,7-Dicyano-1,4-epoxy-1,4-dimethyl-1,4-dihydronaphthalene	80
2.2. Synthesis of dienophilic phthalocyanine	
2.2.1. Asymmetrically substituted mono dienophilic nickel(II)phthalocyanine with AAAB-Symmetry: [2,3,9,10,16,17-Hexa(2-ethylhexyloxy)-23,26-dihydro-23,26-dimethyl-23,26-epoxybenzophthalocyaninato] nickel	81
2.2.2. Symmetrically substituted bis-dienophilic metal free phthalocyanine with ABAB-Symmetry: [2,3,16,17-Tetra(2-ethylhexyloxy)-9,10,23,26-dihydro-9,10,23,26-epoxybenzophthalocyaninato] nickel	82
2.3. Synthesis of Ni/Ni binuclear metallophthalocyanine	83

2.3.1. Pc-tetracyclone adduct :[2,3,9,10,16,17-Hexa(2-ethylhexyloxy)-23,26-dihydro-23,26-dimethyl-23,26-epoxybenzo-24,25-tetracyclone-phthalocyaninato]nickel adduct 83

2.3.2. Synthesis of symmetrically substituted Ni/Ni binuclear metallophthalocyanine 84

2.3.3. Dehydrated binuclear metallophthalocyanine.. 85

2.4. Synthesis of trinuclear metal free phthalocyanine 85

2.4.1. Pc-tetracyclone bisadduct 85

2.4.2. Trinuclear metal free phthalocyanine 86

2.5. Pc-pyridine N-oxide adducts 87

2.6. Synthesis of Phthalocyaninatoindium(III)acetylacetonates 89

2.6.1. [2,3,9,10,16,17,24,25-Octa(2-ethylhexyloxy)phthalocyaninato]indium(III) acetylacteonate 89

2.6.2. [2,(3)-Tetra-*tert.*-butylphthalocyaninato]indium(III)acetylacetonate 89

2.6.3. Symmetrically trinuclear Phthalocyaninatoindium(III)acetylacetonate 90

2.7. Synthesis of Hexadecafluoro(phthalocyaninato)ruthenium(II); $(F_{16}PcRu)_2$ 91

Chapter 6

Literature 92

Abbreviations

br	Broad (NMR, IR)
°C	Centigrade
calcd.	Calculated
cm	Centimeter
$CDCl_3$	Chloroform
δ	Chemical shift
^{13}C CP/MAS	(Cross polarization/magic angle spinning) solid state NMR
d	Doublet (NMR), Day
D	Detector
D A	Diels-Alder reaction
DFWM	Degenerate four-wave mixing
DBU	1,8-Diazabicyclo-[5.4.0]-undec-7-ene
ΔE_e	Energy barrier to injection of electrons
ΔE_h	Energy barrier to injection of holes
dd	Doublet of doublets
DIBAL-H	Diisobutylaluminium hydride
DEPT	Distortionless Enhancement by Polarization Transfer
CH_2Cl_2	Dichloromethane
DMF	Dimethylformamide
EA	Elemental Analysis
E.I	Electron Ionisation
EFISH	Electric field induced second harmonic generation
EHO	2-Ethylhexyloxy
EI	Electron impact
ESA	Excited State Absorption
FAB	Fast Atom Bombardment
FD	Field Desorption
EXAFS	Extended X-ray absorption fine structure spectroscopy
Fig.	Figure
I	Intensity
J	Coupling constant
λ	Wavelength
m	Multiplet (NMR), Medium (IR)

m	Meta position
M	Metal, Molarity (mol l^{-1})
[M$^+$]	Molecular ion (MS)
μ	Magnetic moment
χ$_M$	Magnetic susceptibility
MALDI-TOF	Matrix Assisted Laser Desorption Ionization-time of flight
MHz	Megaherz
m/z	Mass/charge
N	Normal
NIR	Near infrared
NLO	Nonlinear Optics
OL	Optical Limiting
ppm	Parts per million
q	Quartet
RSA	Revere Saturable Absorption
s	Singlet (NMR), Strong (IR)
S	Eye sensitivity curve
S$_0$	Singlet ground state
S$_1$	First excited singlet state
S$_2$	Second excited singlet state
Sh	Shoulder (UV/Vis, IR)
σ	Absorption Cross Section
T$_{Norm}$	Normalised transmission
t	Triplet (NMR)
THG	Third-harmonic generation
TLC	Thin Layer Chromatography
V	Volt
VR	Vibrational relaxation.

I. General part

1 Introduction

Phthalocyanine (Pc) and metallophthalocyanines (MPcs) have been investigated for many years in much detail. They have been mostly used as dyes and catalysts.[1] However, recently, the chemistry of phthalocyanine has been undergoing a renaissance because phthalocyanine and many of its derivatives exhibit properties which are interesting for applications in materials science. Phthalocyanines and structurally related compounds are of interest for example, in nonlinear optics,[2] as liquid crystals,[3] in Langmuir-Blodgett (LB) films,[4, 5] in optical data storage as active materials in recordable compact discs (CDs) and digital versatile discs (DVDs),[6] as electrochromic substance,[7] as semiconducting materials,[5, 8, 9] in rectifying devices,[10] as gas sensors,[11] as photosensitizers [12] and as carrier generation materials in NIR. [13]

The substituted derivatives of phthalocyanines function as acvtive components in various processes driven by visible light: photoredox reactions and photooxidations in solution,[1,14,15] activity in the therapy of cancer,[1,16,17] photoelectrochemical cells,[18, 19] photovoltaic cells,[20, 21] electrophotographic applications. [22, 23]
A summary of some applications are illustrated in the following chart :

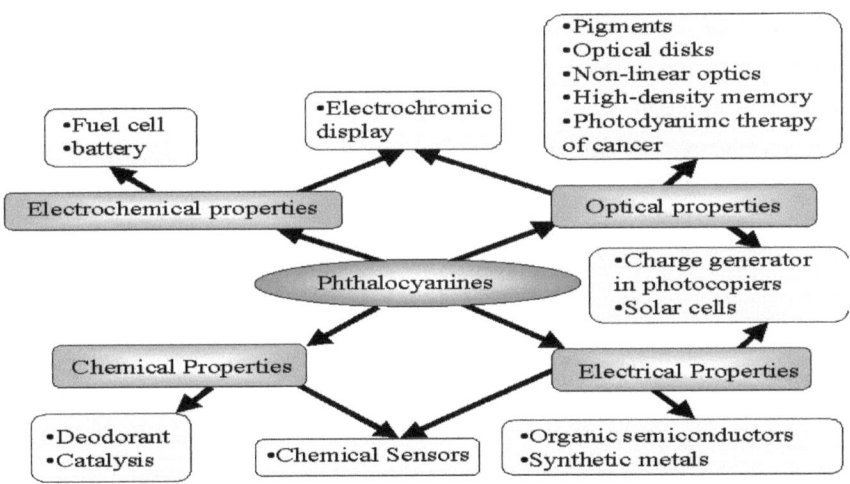

In the following a short introduction to the chemistry and properties of Pc`s is presented.

2. Phthalocyanines [1]

2.1 General Remarks

Interest in phthalocyanines (Pc`s) and their structurally related compounds has grown enormously. The attractive and challenging characteristics of Pc`s are their potential applicability in many fields, their thermal and chemical stabilities, the relative ease with which they can be prepared and purified and the strong dependence of their properties on peripheral and axial substitution patterns.[5] At present, they are a widespread material.

2.2 Discovery of phthalocyanines (Pc`s)

The accidental discovery of phthalocyanine happened more than seventy years ago and a metal-free phthalocyanine was found for the first time in 1907 as by–product during the prepartation of 2-cyanobenzamide. Linstead was the first to investigate the structure of phthalocyanine and describe the name from the Greek words *naphta* (rock oil) and *cyanine* (blue).[24]

2.3 Structures of phthalocyanines

Pc is a conjugated heterocyclic 18-π electron containing compound and is structurally similar to a porphyrin system. Unlike porphyrin, which can be found in the nature such as in hemoglobin, chlorophyll and vitamin B_{12}, Pc`s do not occur in nature. In Pc system, methine-bridges of porphyrin are replaced by aza-bridges and, therefore, Pc`s are tetrabenzotetra-azaporphyrins (Figure 1).

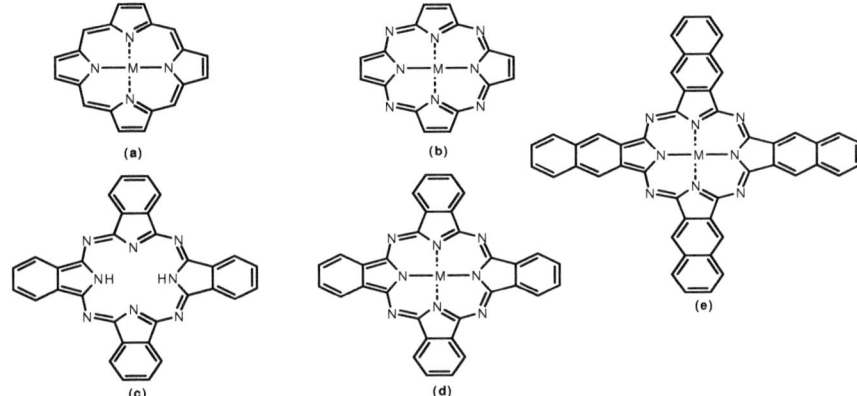

Figure 1: Basic structures of metal-porphyrine (PorM) (a), metal-tetraazaporphyrine (TAPM) (b), metalfree phthalocyanine (PcH$_2$) (c), metal-phthalocyanine (PcM) (d) and metal-2,3-naphthalocyanine (NcM) (e).

Through substitution of hydrogens in metal-free Pc (PcH$_2$) by metal atoms, metallophthalocyanines (PcM) are obtained. Pc complexes with small-sized metal elements, such as Cu^{2+}, Ni^{2+}, Pt^{2+}, in its central cavity form quadratic-planar structures. Larger cations, such as Pb^{2+}, Sn^{2+}, form PcM complexes with a quadratic pyramidal configuration.[25] When large three- or four-fold charged cations, such as Nd^{3+}, Gd^{3+}, Th^{4+}, are introduced, bimolecular complexes in which they are sandwiched between both Pc rings are formed.[26]

2.4 Syntheses of phthalocyanine

2.4.1 General synthetic routes

Up to now over seventy different metallic and non-metallic cations have been incorporated in the central cavity of Pc moiety, thereby enabling the control of the oxidation potential and consequently, the electrical properties of the complexes. The basic synthetic methods of PcM compounds are shown in Scheme 1.

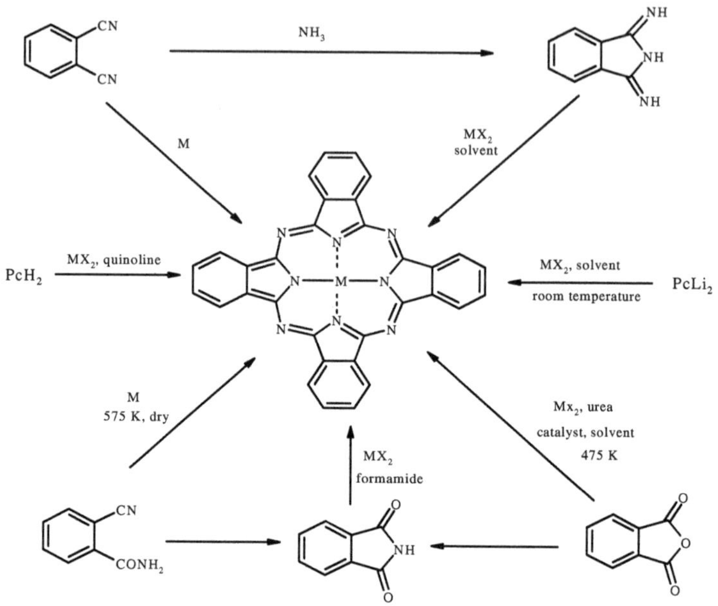

Scheme 1: Basic synthetic methods of PcM.

The majority of PcM can be prepared by the high temperature cyclotetramerization of phthalodinitriles in the presence of corresponding metal or metal salt or by later insertion of the metal into PcH$_2$. On account of the insolubility of unsubstituted Pc in common organic solvent, soluble impurities can be removed by extracting with hot organic solvents or boiling with acids or bases. More soluble substituted Pc`s can be purified by common methods used for organic compounds, usually by chromatography, recrystallization and extraction.

Recently, substituted phthalocyanines are prepared in high yield under microwave heating in the presence of suitable solvent.[27]

Industrially, phthalocyanines were produced by using inexpensive materials like phthalic anhydride and urea [28] which is more useful and cheaper than phthalonitrile route (Scheme 1) to produce higher-volume with lower cost applications as shown in Scheme 2.

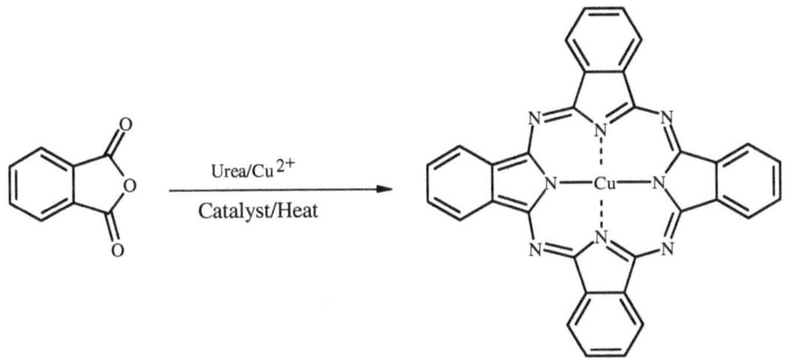

Scheme 2: Industrial preparation of phthalocyanines.

2.4.2 Electrochemical synthesis of phthalocyanines

Yang and co-workers [29] have reported the electrosynthesis of metal free and metallophthalocyanines through the electroreduction of phthalonitrile (PN) in high yields in polar protic and aprotic solvents. Generally, the yields of the electrosynthesis are affected by several factors such as the medium, the reaction temperature, the initial PN concentration, the intensity of the current and the amount of charge which was used during the procedure.

2.5 Soluble phthalocyanines

Due to strong interactions between ring systems, unsubstituted PcMs are practically insoluble in common organic solvents, as mentioned above. The introduction of voluminous hydrophobic substituents into the periphery of the macrocycle enables an increased solubility. Another approach employed is the introduction of axial substituents at the central metal atom decreasing the aggregation effect. By the introduction of selected substituents, the physical and electrical properties of Pc`s can be verified and tailored, resulting in the broadening of their applications.

The best investigated soluble substituted Pc`s are the tetra- and octasubstituted ones.[30] The former generally has higher solubility in common organic solvents than the latter because of its lower degree of order compared with the symmetrical structure of the latter and its higher dipole moment caused by the unsymmetrical rearrangement.

2.5.1 Tetrasubstituted phthalocyanines

Synthesis of tetrasubstituted Pc's can be achieved from 3- and 4-substituted phthalodinitriles, respectively, which results in a mixture of four possible constitutional isomers with different symmetries as shown in Figure 2.

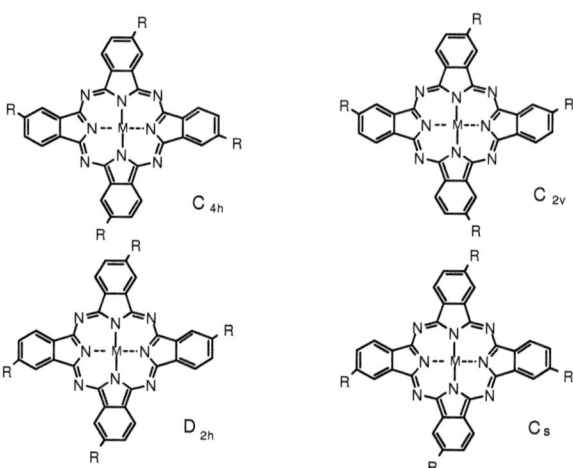

Figure 2: Possible constitutional isomers of tetraubstituted phthalocyanines

The synthetic routes for the preparation of tetrasubstituted phthalocyanines can be summarized as in Scheme 3. The complete separation of the mixture can be carried out by MPLC and HPLC technique which was first achieved by us with the C_{2v} and C_s isomers of (2-Et-$C_6H_{12}O)_4$PcNi.[31] We reported the successful separation of the D_{2h}, C_{2v}, C_s and C_{4h} isomers of tetrasubstituted Pc's and naphthalocyanines.[32]

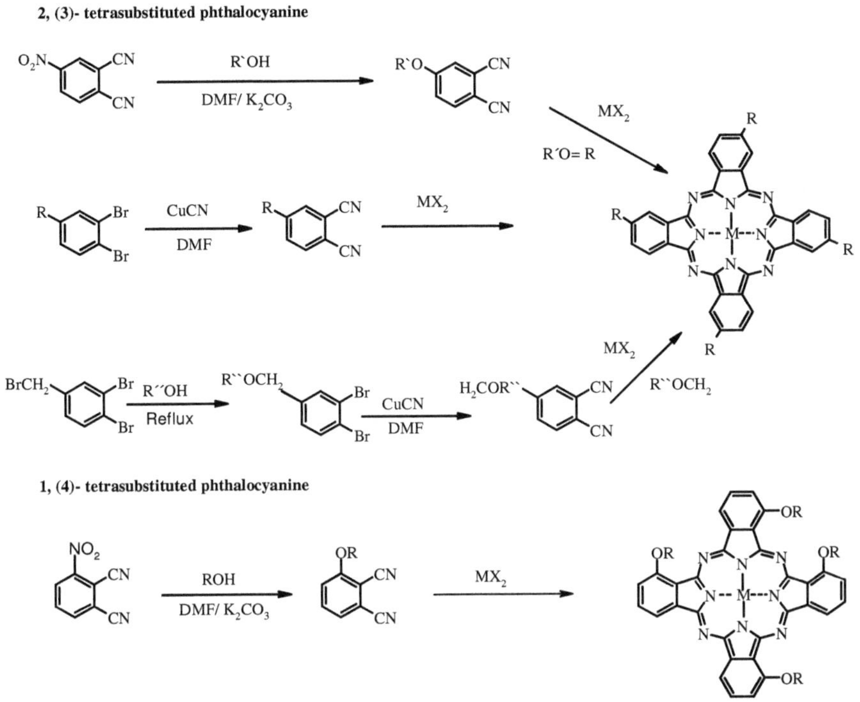

2, (3)- tetrasubstituted phthalocyanine

Scheme 3: Preparation of tetrasubstituted phthalocyanines

2.5.2 Octasubstituted phthalocyanines

The synthetic pathway for octasubstituted Pc`s is similar to that of tetrasubstituted one, only that 3,6- and 4,5-disubstituted phthalodinitriles are used instead of 3- and 4-substituted derivatives, respectively. Two different types of products are obtained. One having substituents in the 1,4,8,11,15,18,22,25- and the other in 2,3,9,10,16,17,23,24-positions, which are depicted as 1,4- and 2,3-octasubstituted Pc`s, respectively. Their structures are represented in Figure 4.

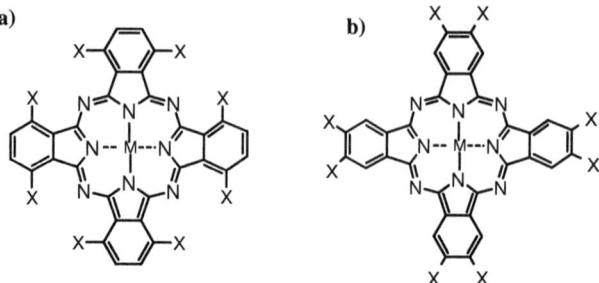

Figure 4: Structure of a) 1,4,8,11,15,18,22,25- and b) 2,3,9,10,16,17,-23,24-octasubstituted phthalocyanines.

Compared to 2,3-substituted macrocycles, the synthesis of 1,4-octasubstituted Pcs is more difficult and usually gives lower yields because of steric hindrance from the substituents. However, they are more soluble in common organic solvents than their isomers due to an out-of-plane arrangement of their substituents.[33]

2.5.3 Unsymmetrically substituted phthalocyanines

This novel kind of Pc`s has attracted interest in recent years not only because of their outstanding electronic and optical properties, but also due to their processability in various applications. For instance, the fabrication of versatile LB thin films,[33, 34] the synthesis of ladder polymers [35] and the preparation of chemically modified electrodes.[36]

Up to now there are three methods available for the synthesis of unsymmetrically substituted Pc`s comprising one different and three identical isoindole subunits (A_3B form), namely the subphthalocyanine approach (Scheme 4), the polymeric support method, and the statistical condensation.[34]

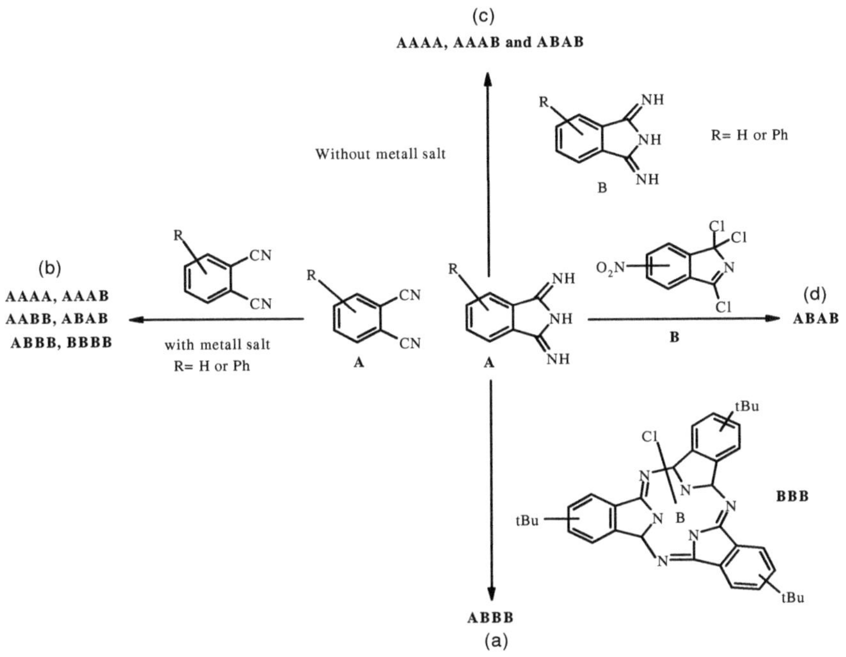

Scheme 4: Overview of the different methods to prepare asymmetric Pc's.

The two pathways (b) and (c) are regarded as selective methods, however, these methods have still many problems, In the case of the subphthalocyanine approach (a), fragmentation of the subphthalocyanine can occur in the reaction process, resulting in all possible Pc's containing all the combination of iminoisonidoline units presented in the starting material. Another method to reduce the number of possible Pc products is the cross condensation (Scheme 4d) of substituted diimnoisoindolines with 6/7-nitro-1,3,3-trichloroisoindolenine. This approach is suitable for the synthesis of bisdienophilic Pcs of the form ABAB.[37] Hence, the most usual method for the preparative work is the statistical condensation between two different phthalodinitriles A and B (Scheme 5) in the appropriate stoichiometric ratio followed by chromatographic separation.[1b]

In a statistical condensation up to six possible Pcs can be formed: Pc molecules containing four A or B units (AAAA or BBBB), three a and one B units (AAAB), two A and two B units (AABB and ABAB), or one A and three B units (ABBB).[31]

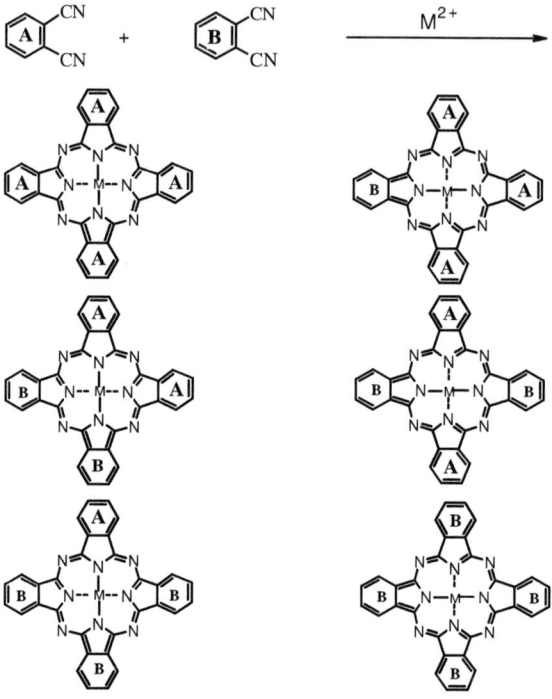

Scheme 5: Statistical condensation and all possible phthalocyanine products.[1b]

Many efforts have been made to synthesize a large number of this kind of Pc derivatives bearing various functional groups.[30]

3 Ladder polymers
3.1 Synthesis of ladder polymers

The synthesis of ladder-type polymers containing macrocyclic units was attempted by Wöhrle and coworkers already in 1968, Various polymers consisting of hemipophyrazine (Hp) subunits were synthesized from tetracyanobenzene and diamines, [38] and other ladder polymers with tetraaza(14)annulene units.[39]

Planar, two-dimensional PcM oligomers and polymers as shown in Figure 5 were prepared from 1,2,4,5-tetracyanobenzene and metal salts, however the corresponding polymeric Pc`s have not been characterize very well.[40]

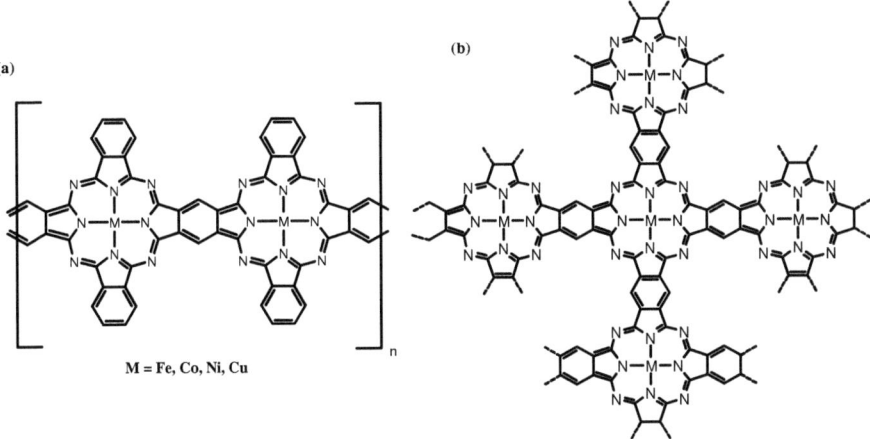

Figure 5: Planar, two-dimensional PcM oligomers

The use of the repetitive DA reaction allows the stepwise formation of macrocyclic ladder oligomers,[41] and the synthesis of polymers of high structural regularity, as shown in Scheme 6.

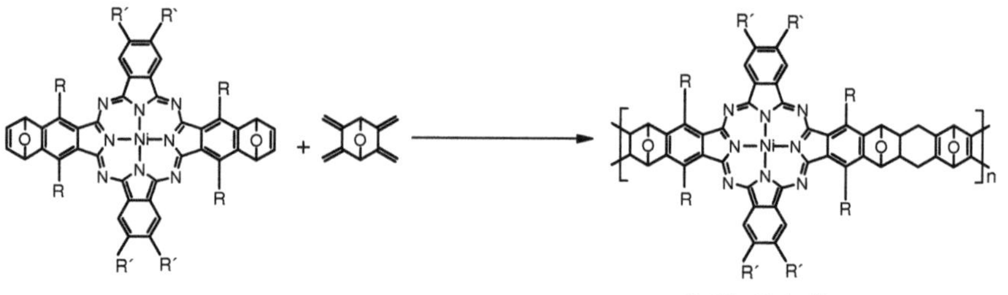

Scheme 6: Possible synthesis of a ladder polymer with macrocyclic subunits in a repetitive DA reaction.

3.2 Selected dimers, trimers and oligomers of phthalocyanines

Pc dimers, trimers and oligomers were reported by our group.[35,42] The Pc dimer (Figure 6a) was prepared from a precursor obtained from three phthalonitrile units and one 6,7-dicyano-1,4-dihydro-1,4-epoxynaphthalene. To this epoxy unit, tetraphenylcyclone was fused by a Diels–Alder reaction, and the resultant Pc derivative was reacted with π-benzoquinone in refluxing toluene (Figure 6a).

Another example of a linear ladder-type Pc trimer (Figure 6b) was also synthesised by our group from Pc's substituted with two 1,4-epoxy units at opposite positions applying the Diels-Alder reaction.[42]

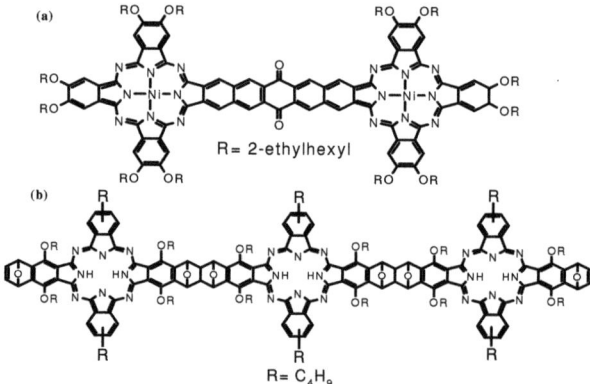

Figure 6: Planar Pc-dimer (a) and trimer (b)

4 Stacked phthalocyanines

4.1 Nonbridged stacked polyphthalocyanines

Phthalocyanines crystallize in a columnar arrangement through a large variety of metal-macrocycle-ligand combinations as shown in Figure 7a,b.

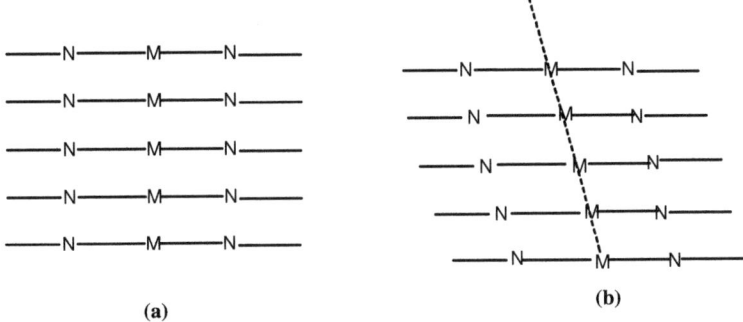

Figure 7: Possible stacked arrangements of metal-macrocycle complexes: (a) one dimensional arrangement and (b) α- or β-crystal structures.

Nonbridged stacked polyphthalocyanines could be classified intro two categories: unsubstituted [43] and substituted phthalocyanines.[44] Some examples are given in Figure 8a,b.

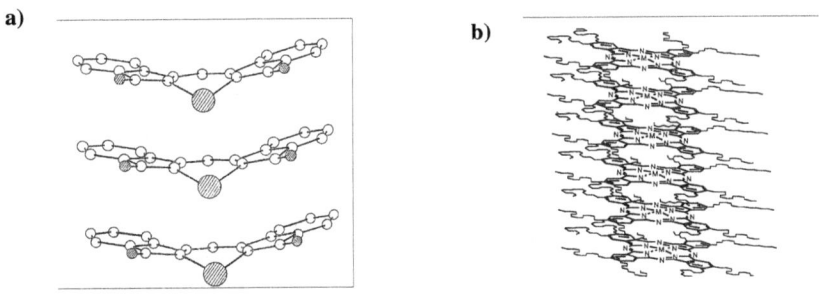

Figure 8: a) Columnar structure of unsubstituted PcPb, b) Columnar structure of octasubstituted metal-phthalocyanine.

4.2 Bridged stacked polyphthalocyanines

The stacked arrangement can also be achieved by bidentate ligands L (Figure 9).[8, 30, 43] These columnar structures represents a noteworthy achievement in materials science.

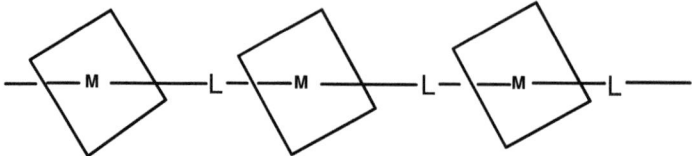

Figure 9: General structure of a metal-macrocycle complex in the stacked arrangement with the central atoms connected via a bidentate ligand L. The square indicates the coordinating macrocycle.

This group of stacked polymeric phthalocyanines is characterized by the presences of coordination bonds as links which bridge the metal-macrocycles adducts via the ligand. The central metal atoms are in divalent, trivalent or tetravalent oxidation states as shown in Scheme 7. [30]

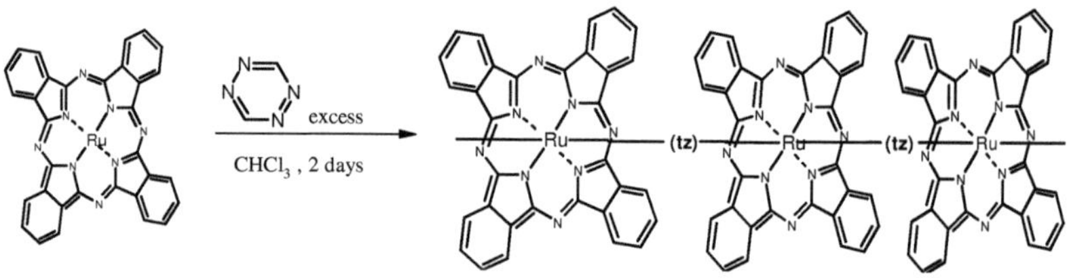

Scheme 7: Synthesis of [PcRu(tz)$_n$]

The cofacially linked stacked polyphthalocyaninatometalloxanes [PcMO]$_n$ (M= Si, Ge, Sn) are shown schematically in Figure 10.[46, 47]

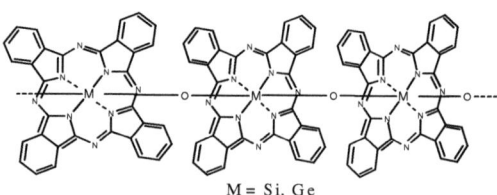

M = Si, Ge

Figure 10: µ-Oxo-phthalocyaninatometal(IV) compounds

5 Optical limiting: A nonlinear optical effect

Nonlinear optical (NLO) phenomena are associated with changes in the optical properties of materials which are exposed to intense light. The importance of the study of NLO phenomena relies mostly upon the manifold possibilities of their exploitation in the development of photonic technology. Infact, several optical devices or technologies such as optical rectifiers, optical switches, dynamic holography and optical data-recording are based on NLO effects. [48]

One of the first observations of nonlinear optical phenomena was reported by Franken et al.,[49] when frequency doubling in a quartz crystal illuminated by a laser occurred.

In general, light intensity modifies the absorptive, refractive and scattering properties of the illuminated system once $I > I_{lim}$. (see section 5.2). In the case of molecular species, the extinction coefficient k can vary with the intensity I according to the relationship:

$$k = k_0 \,[1/\,1+(\,I/I_{lim})]\;(1)$$

with k_0 corresponding to the low intensity limit value of k. Equation (1) describes the optical behaviour of a saturable absorber and expresses the fact that the extinction coefficient k decreases with the increase of the incident intensity. In doing so, the optical system gets more transparent at higher incident intensities, and behaves like an intensity-activated optical switch. On the other hand, in the case of optical limiting (OL) systems the opposite situation is verified, i.e. the optical system has an extinction coefficient which increases reversibly with the augmentation of the incident radiation intensity.[50, 51] This phenomenon constitutes the so called reverse saturable absorption (RSA), and takes place mostly with the irradiation of organic dyes, donor-acceptor molecules, fullerenes and in less extent inorganic semiconductors. The OL effect can be produced also by means of several other mechanisms based on fundamental optical processes different to absorption, e.g. refraction and/or scattering. In fact, OL effect based on radiation diffraction or scattering does not allow the formation of a well-resolved image once the incident light rays have interacted with the optical limiter system. As a consequence, optical limiters based on phenomena other than absorption have the practical limitation of not being useful for the protection of those complex light-sensitive elements, e.g. the eye, employed in direct viewing operations which require a clear vision of the surrounding environment. For this reason among the available systems for the limiting of intense radiations, those based upon the phenomenon of RSA absorption are preferable.

5.1 Optical limiting phenomena

Optical limiters are an important application of the nonlinear optical properties of the phthalocyanines. An optical limiter is a device that has a high transmission at normal light intensities and a decreasing transmission for intense beams (Figure11). Such devices can perform useful optical functions; the most obvious is the protection of human eyes from high-intensity light sources.[52] Various designs for optical limiters have been described [53] based on several mechanisms giving rise to nonlinear optical responses.[54, 55]

5.2 Nonlinearity of the optical effect

Devices such as optical limiters or saturable filters show the OL effect when the intensity of a light beam is strongly attenuated once the input intensity exceeds a threshold value (I_{lim}).

The latter is determined by the characteristics of the system interacting with the light beam, and represents a critical parameter for the evaluation of the OL properties of the device. The ideal behaviour of an optical limiter is shown in Figure 11, I_{out} and I_{in} being respectively, the intensity of the light beam transmitted by the optical limiter and the intensity of the incoming light. [56]

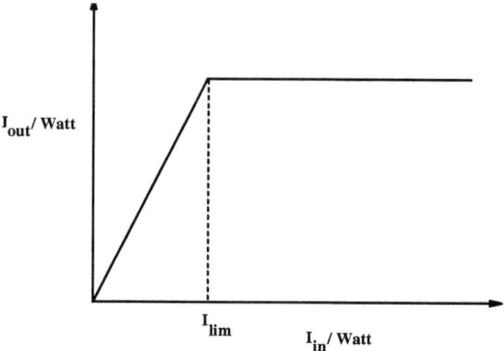

Figure 11: Trend of light intensity I_{out} transmitted by an ideal optical limiter vs. incoming light intensity I_{in}. The threshold intensity I_{lim} at which I_{out} saturates is indicated.

The transmittance **T** [= d(I_{out}) / d(I_{in})] of an ideal optical limiter is not constant within the whole regime of irradiation and **T** becomes a function of I_{in} with **T₀** for $I_{in} \gg I_{lim}$. A useful parameter for the evaluation of the OL effectiveness of different systems is the intensity threshold defined as the incident intensity value at which the transmittance of the system is equal to the 50% of the linear transmittance.

An important mechanism for the occurrence of NLO effects like OL is optical pumping. In this case the incident laser frequency approaches a transition frequency in the molecule. The light is absorbed, causing transitions to the excited state. The optical properties of the excited state differ considerably from those of the ground state and the higher the population in the excited state, the larger the changes in the optical properties of the material.

Optical pumping involves real transitions to the excited state, and this phenomenon is quite different from the small perturbations of the electronic cloud which are verified in the regime of linear polarization. Optical pumping can induce both saturable and reverse saturable absorption depending on the difference of the absorption properties of the system between the ground and excited states at the wavelength of irradiation.

5.3 Optical limiting: mechanism and model

The intensity dependent absorption coefficient has been derived assuming that Reverse Saturable Absorption is the dominant dissipating mechanism. The dynamic solution to the optical limiting problem has been derived and will be outlined briefly here. In general the electronic states can be described via a 5 level energy diagram (Figure 12).

The system can be described using the following series of rate equations where the population of the levels S_0, S_1 and T_1 are given by n_1, n_2 and n_3, respectively. S_i represents singlet levels and T_i represent triplet levels. τ_{isc} and τ_{ph} are the inter-system crossing and phosphorescence lifetimes, respectively.

$$\frac{\partial n_1}{\partial t} = -\frac{\sigma_0 I}{h\nu} n_1 + \frac{n_2}{\tau_{10}} + \frac{n_3}{\tau_{ph}}$$

$$\frac{\partial n_2}{\partial t} = \frac{\sigma_0 I}{h\nu} n_1 - \frac{n_2}{\tau_{isc}} - \frac{n_2}{\tau_{10}}$$

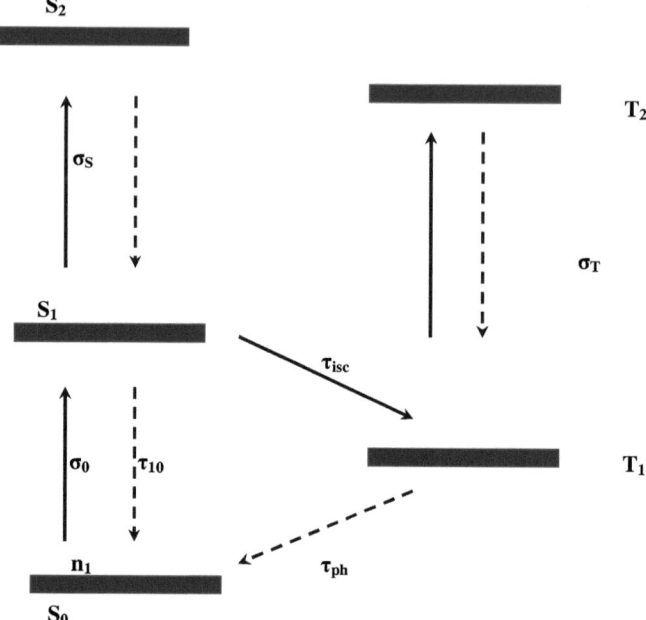

Figure 12. Generalized 5 level model used in calculation of excited state dynamics of phthalocyanine system. S_i represents singlet levels and T_i represents triplet levels. Solid arrows imply an excitation resulting from photon absorption and dashed arrows represent relaxations.

The attenuation of the laser beam is governed by the steady state approximation, which is valid when the pulse width is much longer than any of the relaxation times, all the time derivatives may be set equal to zero. This is valid for nanosecond pulses as the lifetimes in phyhalocyanines are typically in the order picoseconds. The total population, N is now introduced and n_i is expressed in terms of it.

$$n_1 = N - n_2 - n_3$$

This can be substituted back into the rate equations above to give,

$$n_2 = \frac{\sigma_0 I}{h\nu}\left(N - n_2 - n_2\frac{\tau_{ph}}{\tau_{isc}}\right)\tau_{isc}$$

where the following approximation has been applied

$$\frac{1}{\tau_{isc}} \approx \frac{1}{\tau_{isc}} + \frac{1}{\tau_{10}}$$

Noting that $\tau_{isc} \ll \tau_{ph}$ and using the expression

$$I_{Sat} = \frac{h\nu}{\sigma_0 \tau_{ph}}$$

an expression for n_2 in terms of the total population N is found and is given by

$$n_2 = \frac{\tau_{isc}}{\tau_{ph}} \cdot \frac{I}{I_{Sat}} \cdot \frac{N}{1+\frac{I}{I_{Sat}}}$$

n_3 is then found to be given by

$$n_3 = \frac{I}{I_{Sat}} \cdot \frac{N}{1+\frac{I}{I_{Sat}}}$$

and n_1 is then expressed as

$$n_1 = \frac{N}{1+\frac{I}{I_{Sat}}}$$

The populations for each level, can then be substituted back to give

$$\alpha(I) = \frac{\sigma_0 N}{1+\frac{I}{I_{Sat}}}\left(1+\frac{\sigma_S}{\sigma_T} \cdot \frac{\tau_{isc}}{\tau_{ph}} \cdot \frac{I}{I_{Sat}} + \frac{\sigma_T}{\sigma_0} \cdot \frac{I}{I_{Sat}}\right)$$

The merit coefficient κ can be defined as $\kappa = \sigma_T/\sigma_0$, and as the triplet yield is ~ 100% for phthalocyanines and $\tau_{isc} \ll \tau_{ph}$, we are then left with

$$\alpha(I, I_{Sat}, \kappa) = \frac{\alpha_0}{1+\frac{I}{I_{Sat}}}\left(1+\kappa \frac{I}{I_{Sat}}\right)$$

which can be fitted to the data. The above equation was applied to the optical limiting data of all compounds.

The excited state absorption is not the sole mechanism with which optical limiting effects can be achieved. In fact other mechanisms can intervene in processes of optical limiting like thermal refractive beam spreading,[57] non-linear refraction[58] or optical breakdown-induced scattering.[59, 60] In the latter case, a perturbation of the electronic distribution with the electric field associated with the incident light, or a directional rearrangement of polar molecules with intensity-dependent changes of the refractive index represent the active mechanisms for the occurrence of optical limiting.

In the following, paragraphs the nonlinear optical properties of phthalocyanines will be explained.

5.4 Phthalocyanines for optical limiting

Phthalocyanines and their derivatives are an important group of molecular materials for NLO applications.[61] The reason for this is ascribed to the high delocalization of the electronic cloud between the macrocyclic ligand and the coordinated metal atom, which affords large NLO coefficients.[62]

Most optically linear active phthalocyanines which have been reported in the literature are classified into three groups [63]: (i) Pc`s containg carbons in the side chain. This type of Pc is further divided into two gropus, i.e. Pc`s with eight long alkyl chains attached directly or indirectly onto the periphery of the pc core,[64] and pc`s linked with small cyclic compounds,[65] (ii) Pcs with optically active aromatic molecules. This group includes helicene and binaphthyl-substituted species.[66] and (iii) phthalocyanines with an optically active aromatic molecule as peripheral substitution or as an axial ligand. [67]

To be used as optical limiters, phthalocyanines must show a high level of linear transmission and large nonlinear absorption over a broad spectral bandwidth, as well as a high threshold for damage. Moreover, the nonlinear absorption must appear within a subnanosecond response time. Among the phthalocyanine based nonlinear absorbers that have been used as optical limiting materials and approach the necessary characteristics for a practical device one example called: (state of art molecule) can be cited: $(tBu)_4PcInX$ (compounds **A,** [68a] Figure 13).

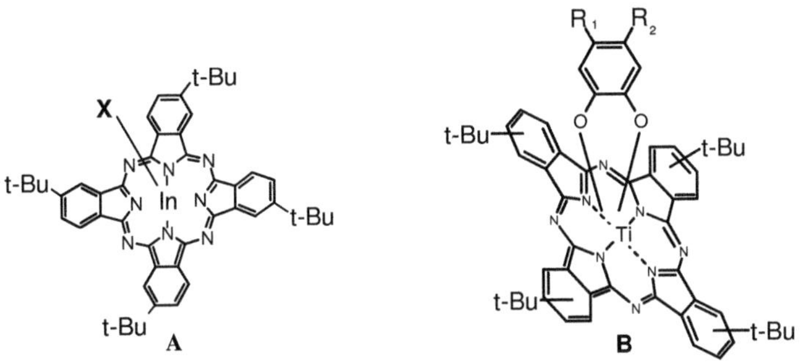

X= CL or Ph or *p* -TMP
(Mirxture of isomers)

R_1= t-Bu, R_2= H

Figure13: Chloro-and arylindium(III)phthalocyanines **A**, Phthalocyaninatotitanium(IV)catecholato complex **B**

Another significant factor influencing the OL properities of Pc`s is the variation of the axial substituents. In phthalocyaninatotitanium(IV) catecholato complexes **B** the charateristic pattern of the nonlinear transmission curves depend on the nature of the various axial substituents (compounds **B**, Figure 13).[68b]

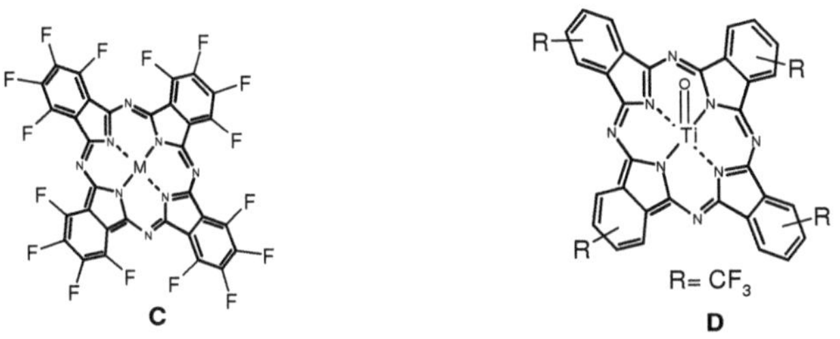

Figure14 : Fluorinated substituted Pc`s **C**, Oxo-titanium Pc`s **D**

The introduction and variation of peripheral substituents in phthalocyanines can also modify the structural arrangement of the molecule or the spatial relationship between neighbouring molecules. The use of unsymmetrically substituted phthalocyanines is another approach to enhance the OL properties of phthalocyanines, since the introduction of nonsymmetry can change the electronic structure of the macrocycle.

Peripheral electron withdrawing or relasing groups also change the electronic structure of the substituted Pc`s, as e.g. shown in Figure 14 for compounds **C** and **D**. The presence of electron withdrawing substitutents like fluoro on the Pc ring, as in compound **C** or peripheral substituents, as in compound **D**, enhances the OL effect due to the increase of the transition dipole moment between the excited states involved in the electronic transition responsible for the OL effect.[68b]

The optical limiting property of sandwich-type rare earth metal diphthalocyanines, Eu[Pc(R)$_8$]$_2$[R= n-C$_7$H$_{15}$, OC$_5$H$_{11}$] (Figure 15), has been investigated via fluence-dependent transmittance measurments.[69] The results demonstrated that Eu[Pc(OC$_5$H$_{11}$)$_8$]$_2$ **E** exhibited better optical limiting behavior than Eu[Pc(C$_7$H$_{15}$)$_8$]$_2$ **F**. This attributed to the enhanced delocalization of the π–conjugated system in Eu[Pc(OC$_5$H$_{11}$)$_8$]$_2$ because of the stronger electron-donating ability of alkoxy relative to that of alkyl.

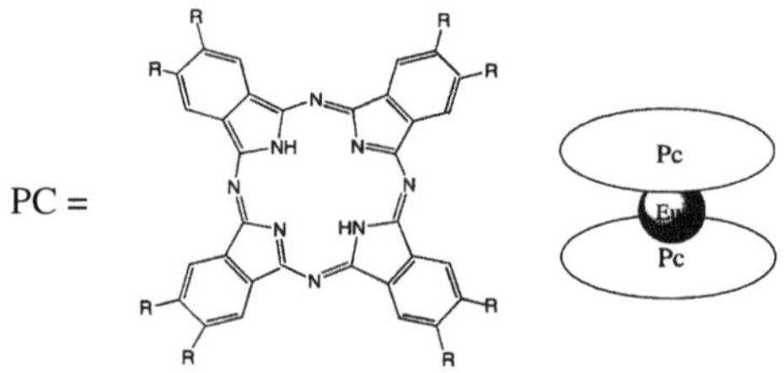

E, R = OC$_5$H$_{11}$

F, R = n-C$_7$H$_{15}$

Figure 15: Structure of sandwich-type rare earth metal dimers Pc`s **E** and **F**

We have recently published the synthesis of several pc`s dimers [70, 71] connected by an oxo-bridge. Indium and gallium phthalocyanine dimers, **G** and **H**, are two example (Figure16), synthesized by the reaction of the corresponding chloro complexes with concentrated H$_2$SO$_4$ at –20 °C. These compounds have good solubility in common organic solvents and show no aggregation in solution form. The transient absorption spectra revealed their absorption band centered at ~ 520 nm, in agreement with their good optical limiting performance at 532 nm.

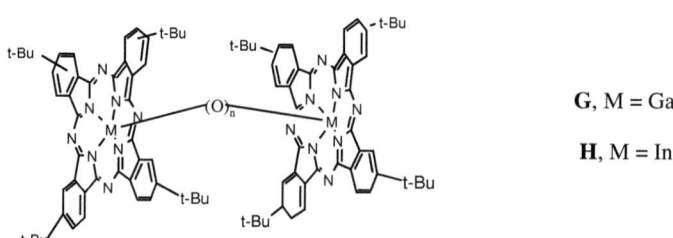

G, M = Ga

H, M = In

Figure 16: Structure of metal-metal oxo-bridged dimer Pc`s **G** and **H**

A few metal-metal directly linked dimers have been reported over the last few years [72-76] but no optical data were recorded. A new axially linked indium phthalocyanine dimer with a direct In-In bond [tBu$_4$PcIn]$_2$ 2tmed (temd = N,N,N`,N`-tetramethylenediamine) **I** (Figure 17) has been synthesized by our group.[77] The values of Im{$\chi^{(3)}$} and γ at 532 nm for **I** were determined to be $(0.9 \pm 0.2) \times 10^{-11}$ and $(1.45 \pm 0.3) \times 10^{-32}$ esu, respectively. Its optical limiting response was significantly enhanced by dimerization. [78]

We have reported a pc dimer assembled by two Ti atoms (Figure 17) synthesized by the dimerization of titanium(IV) phthalocyanines through the bridging species tetrahydroxy-p-benzenoquinone. The dimeric complex (tBu$_4$PcTi)$_2$O$_4$(C$_6$O$_2$) **K** shows an improvement of the nonlinear optical behavior with respect to the monomer.[79]

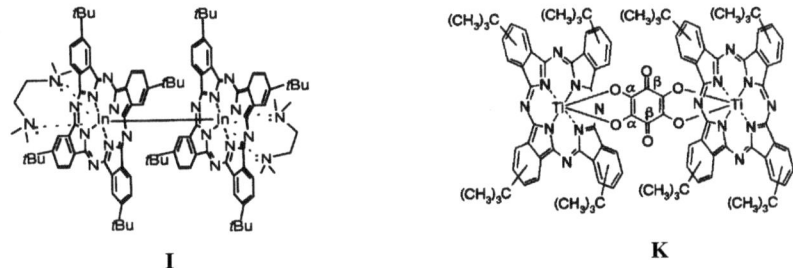

Figure 17: Structure of metal-metal directly linked dimer Pc **I** and bridged titanium(IV) phthalocyanine **K**

II. Aim of the work

In the present work several aspects of phthalocyanine chemistry have been planned for investigation:

1. As a continuation of an earlier non successful approach we plan to synthesize for the first time a fully conjugated soluble phthalocyanine dimer of structure **11** as represented in Scheme 8.[82a]

The rather complicated synthesis of **11** starting with the unsymmetrical phthalocyanine **4a** is described on pages 29-37.

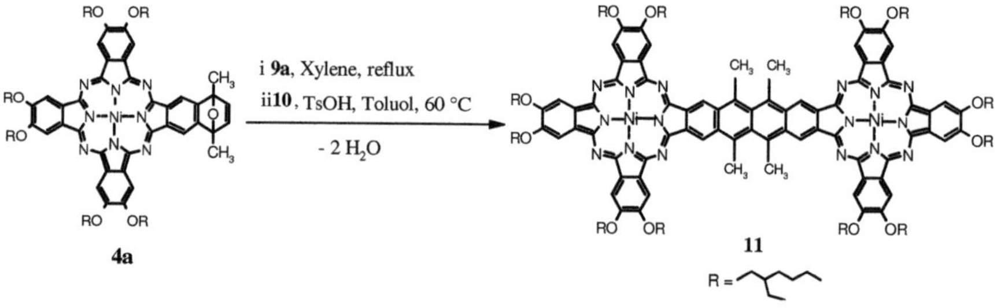

Scheme 8: Symmterically substituted binuclear nickel phthalocyanine **11**

2. Compound **4a** for the first time it should be reacted with heterocyclic pyridine-*N*-oxides **14a-c** via cycloaddition to form the correspounding furopyridine-type phthalocyanines **15a-c** (which will be applied to several aspects in pharmaceutical industries) as shown in Scheme 17. (see page 46).

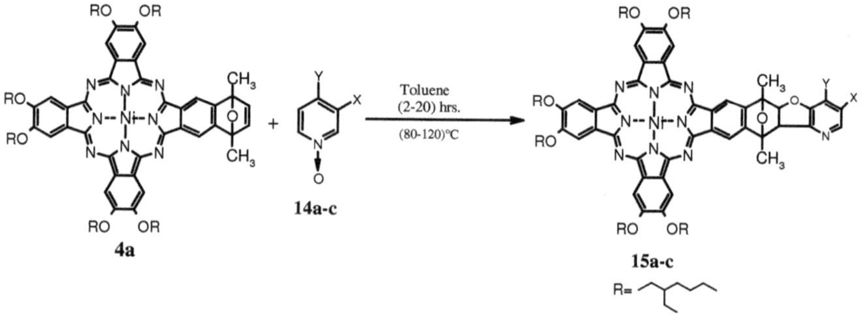

Scheme 9: Synthesis of furopyridine-type phthalocyanines **15a-c**

3. The extensive investigations which have been done in our group on the optical limiting properties of axially substituted indium phthalocyanines will be extended to a new type of compounds namely the axially substituted phthalocyaninatoindium(III) acetylacetonates **16, 19** and the phthalocyaninatoindium(III)acetylacetonate trimer **20** (Figure 18). The syntheses had to be developed for **16, 19** and **20** which is described in Schemes 18-20 on pages (51-54), respectively. The optical limiting properties of **16, 19** and **20** will be compared with other axially substituted phthalocyanine indium compounds e.g. with the ``state of art`` molecule, tetra-*tert*-butylphthalocyaninatoindium(III)chloride (**21a**).

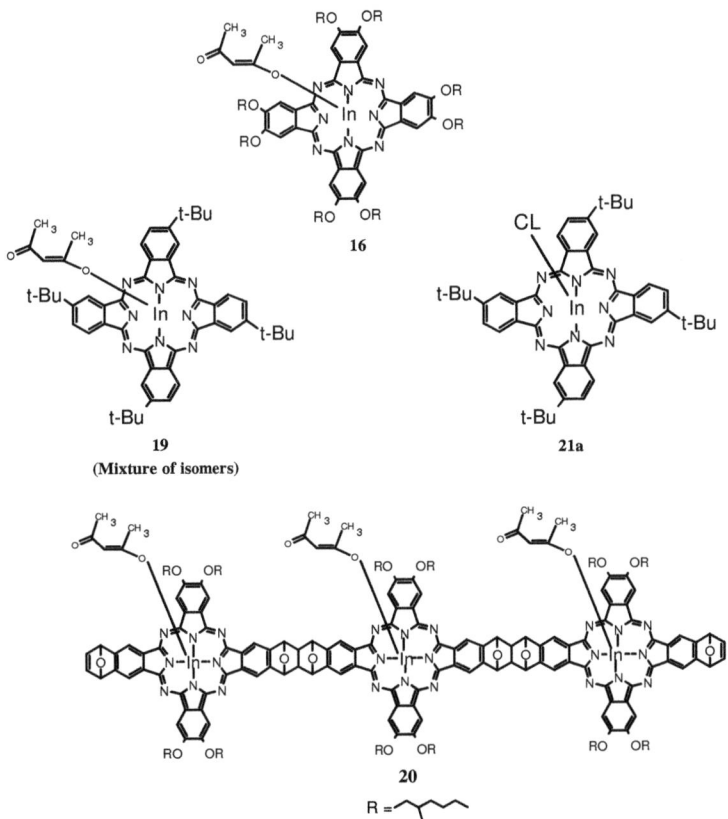

Figure 18: Phthalocyaninatoindium(III)acetylacetonates **16, 19** and **20**

4. For the first time hexadecafluoro(phthalocyaninato)ruthenium(II) F_{16}PcRu **23a,b** will be synthesized and structurally characterized e.g. with EXAFS spectroscopy. Pc **23** will be compared in its general properties with the unsubstituted PcRu especially concerning the question whether or not it has monomeric (**a**) or a dimeric structure (**b**) which is described in Scheme 21 on page 66.

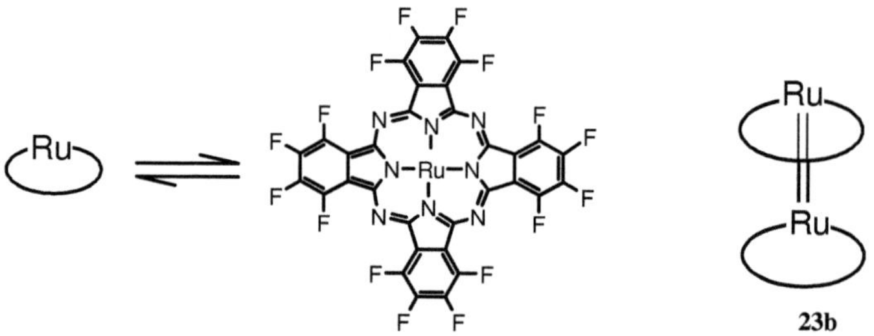

Figure 19: Hexadecafluoro(phthalocyaninato)ruthenium(II) **23a,b**

III. Results and Discussion

1 Synthesis of the fully conjugated soluble binuclear phthalocyanine 11

A binuclear nickel Pc **L** (Scheme 10) with twelve (2-ethylhexyloxy) substituents was earlier described in a thesis. [80] To prepare a dehydrated binuclear nickel Pc was not successful [80] only a binuclear nickel Pc with the structure **M** shown in Scheme 10 containing a hydroxyl group was formed after dehydration of **L** with *p*-toluenesulfonic acid. [80]

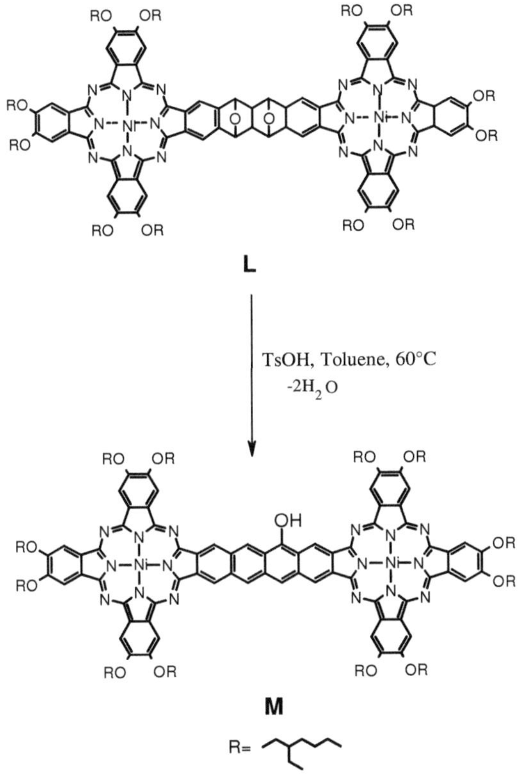

Scheme 10: Synthesis of binuclear phthalocyanines [80]

To avoid the formation of a hydroxylated product **M** in our new attempt for the synthesis of a fully conjugated system methyl groups were introduced into the structure to obtain as the final the dimer **11** as shown in page 37.

The synthesis of **11** (page 37) starts with new phthalonitriles, one of them carring two methyl groups on the epoxy ring **2B** to afford the new unsymmetrically dienophilic PcNi type **AAAB 4a** by the statistical condensation shown in Scheme 11.

1.1 Synthesis of the dienophilic phthalocyanines

To date, the best method to obtain conjugated binuclear and trinuclear metallophthalocyanines with high structural regularity is the repetitive Diels–Alder reaction. [80] The monomers used for such a reaction also allow the stepwise synthesis of defined oligomers. The starting Pc`s have to fulfill three requirements to qualify as substrates for the synthesis of these Pc`s. They must (i) possess Diels-Alder functionalities, (ii) exhibit AAAB or ABAB symmetry, and (iii) have a conjugated π-electron system. [81]

The complicated synthesis of binuclear Pc **11** is demonstrated in Schemes 11,12, 13 and 14 on pages (29, 32, 34 and 37).[82]

General Synthesis

The syntheses described in Schemes 12-14 started with the preparation of the precursor **4a** which is obtained from the statistical condensation between **1A** and **2B** in Scheme 11 resulting in the formation of six different phthalocyanines, viz. **AAAA** (self-condensation of **1**), **AAAB**, **AABB**, **ABAB**, **ABBB** and **BBBB** (self-condensation of **2**) (c.f. Scheme 11). By changing the ratio between the two dinitriles **1** and **2** in the statistical synthesis, the resulting amount of each isomer can be varied, so that the required isomer **4a,b** is obtained in good yield when the ratio between the dinitriles **A** (**1**) and **B** (**2**) is 3: 1.

The separation of these products by common chromatographic methods is not easy due to their tendency to form aggregates. One method of reducing the number of possible combinations is by attaching bulky alkoxy substituents such as (2-ethylhexyloxy) at the 3,6 positions of the reacting component **1** which confers good solubility and suppress aggregation tendencies. Diels–Alder monomers based on unsymmetrical nickel hemiporphyrazines and phthalocyanines that can be used as dienes and dienophiles have previously been described.[83-85]

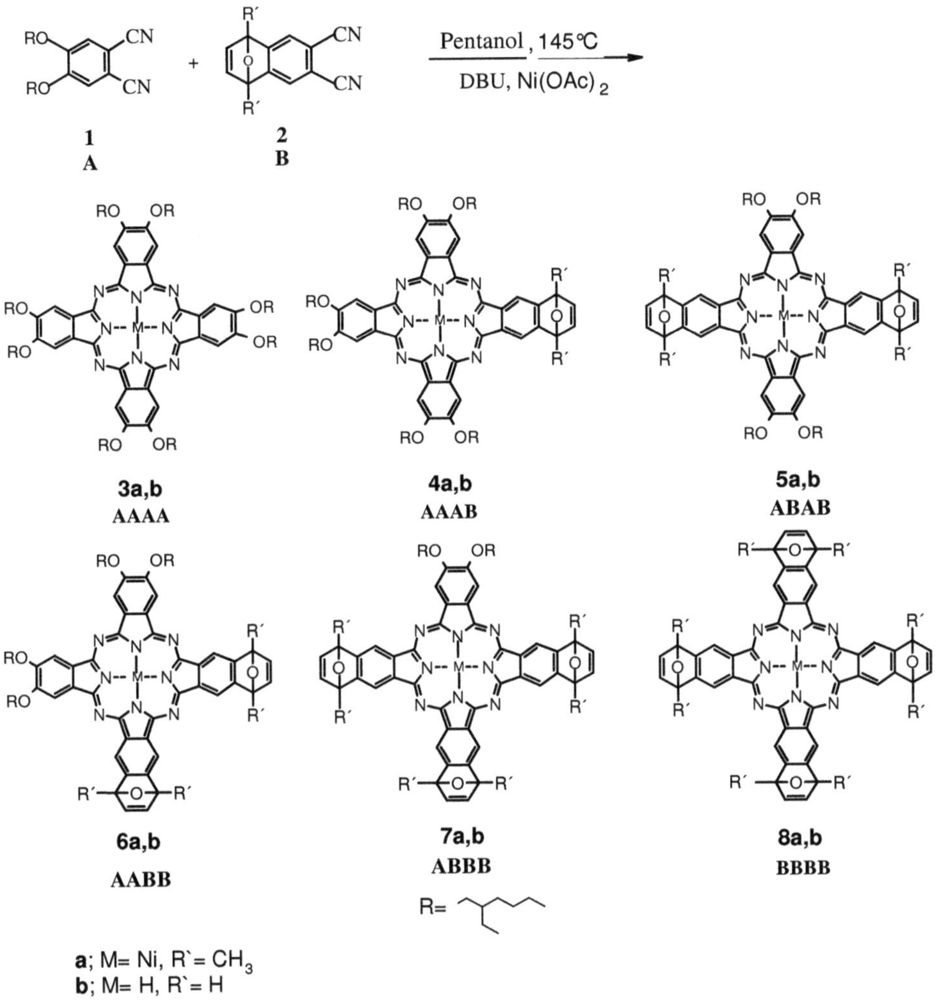

Scheme 11: Synthesis of phthalocyanines **3–8**

In Table 1 an example with the number of permutations and the relative portions of products **3–8** for different stoichiometry is given.[80, 86]

Table 1: Expected relative portions from the statistical condensation mixture of products.

A:B	3 (AAAA)	4 (AAAB)	5 (ABAB)	6 (AABB)	7(ABBB)	8 (BBBB)
1:1	6.25	25	12.5	25	25	6.25
3:1	31.6	42.2	7.0	14.1	4.7	0.4
9:1	65.6	29.2	1.6	3.2	0.4	0.01

1.2 Synthesis of asymmetrically substituted mono dienophilic nickel(II)phthalocyanine 4a with AAAB-Symmetry:

The preparation of the asymmetrically alkoxy substituted PcNi **4a (AAAB)** was realized by condensation reaction between one equivalent of 6,7-dicyano-1,4-dihydro-1,4-epoxy-1,4-dimethylnaphthalene [87] **(2)**, 3.5 equivalents of 1,2-dicyano-4,5-bis(2-ethylhexyloxy)benzene [88] **(1)** and nickel(II) acetate in pentanol at 145 °C in the presence of catalytic amounts of DBU [1,8-Diazabicyclo(5.4.0)-undec-7-ene]. In this process the main product was **4a**. The Pc`s **3a** and **8a**, respectively, are formed due to self-condensation of **1** or **2** in 12 % yield, (Scheme 11) They were also isolated during the chromatographic workup. The blue green compound **4a**, was separated from the first fraction containing **3a** with dichloromethane (DCM) as eluent. After the collection of the first fraction, the second, which was the desired fraction, was eluted (compound **4a**), using a mixture of DCM– hexane (2:1). Chromatography was continued, but the subsequent fractions were eluted all together with a mixture of DCM: ethyl acetate (4:1), since the rest of the products were not needed for our purpose (see Experimental part).

1.2.1 Spectroscopic characterization of phthalocyanine 4a

The ^{1}H-NMR spectrum for **4a** shows the expected aggregation and broadening of the signals. The multiplets of the 2-ethylhexyloxy substituents present an unstructured signal between 1.02 and 2.09 ppm. The aromatic signals, in spite of the broadening, appear as singlets. Characteristic signals for the epoxybenzo unit in **4a** are two singlets arising from 1-H and 5-H.at 7.39 and 9.09 ppm respectively. In addition, strong and well resolved resonances are observed for the methyl groups on the epoxy ring at $\delta = 1.2$. The methyl groups increase the solubility and stability properties of Pc **4a**. (see Scheme 11, compound **4a**).

In the ^{13}C-NMR spectrum of **4a,** there are characteristic signals for the epoxybenzo units at 144.4 ppm (C-1). For C-11 there is a resonance at 153 ppm and for C-5 and C-10 appear at 102.5 and 112.6, respectively, (see experimental, page 79).

The use of IR spectroscopy is of restricted value in the characterization of phthalocyanines. The characteristic vibrations are predominantly found in the fingerprint region. Comparing these vibrations in a new complex with those of an authentic phthalocyanine sample was the technique formerly used as proof for the formation of a new phthalocyanine derivative. It was shown that this technique was useful to determine the particular polymorph of a solid phthalocyanine.[89] This is especially important in solid state applications, where mostly unsubstituted phthalocyanines are employed.

However, in the case of substituted phthalocyanines, the typical vibrations of the substituents dominate the spectrum.

In IR spectrum, the aliphatic band of C-H appear between 3000 and 2800 cm^{-1}.

In the UV/Vis spectra of phthalocyanines, besides the intense Q-bands (601-666 nm), the Soret-band appears between 310 and 400 nm. Both bands are due to π - π* transitions in the 18-π–electron systems of the macrocycle.[90-92]

1.3 Synthesis and spectroscopic characterization of Pc-tetracyclone adduct 9

Compound **4a** was converted into the corresponding Pc-tetracyclone adduct **9** by heating the corresponding Pc **4a** with 1 equivalent of tetraphenylcyclopentadien-1-one (tetracyclone) in toluene at 75 °C (3–4 days) in 70 % yield after chromatographic work up.

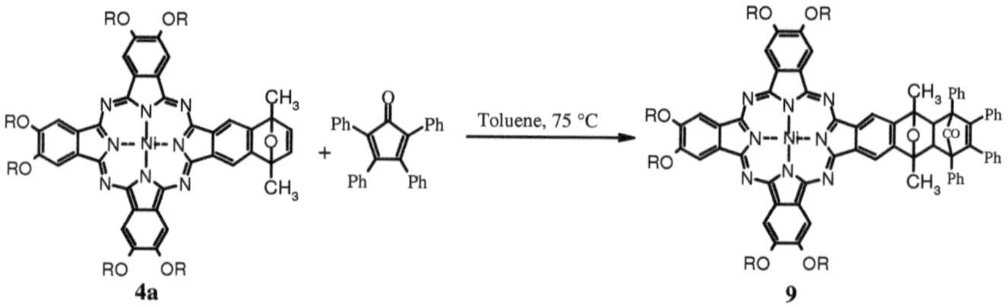

Scheme 12: Synthesis of Pc-tetracyclone adduct **9**

By addition of tetracyclone to the isolated double bond in **4a**, the ^{1}H NMR spectrum of **9** shows the characteristic signals from the phenyl protons in the aromatic region between 7.2 and 7.68 ppm. The signals of the vinylic protons 1-H at $\delta = 7.00$, τ for **4a** disappear and singlets for the methine protons 3-H are found at $\delta = 3.39$ for **9**, which is in agreement with the expected *exo*-orientation of the tetracyclone adduct. The alkyloxy chains in the nonaromatic region resonate between 1.10 and 2.09 ppm.

The ^{13}C-NMR shows also the characteristic C=O signal at 196.6 ppm, among the other assigned signals (see experimental, page 81).

The UV/Vis spectrum of **9**, when compared to its precursor, **4a**, shows a seven nm red-shift, from 666.0 nm in **4a** to 673.0 nm, for the Q-band. Also the appearance of the carbonyl bands in the IR spectrum for **9** at $\nu = 1716$, is in agreement with the structure of the tetracyclone adduct.

In the FD-MS spectrum, two fragments peaks at *m/z* = 1819.9 and 1408.5 (M$^+$ − isobenzofuran) for **9a** are seen. These peaks are due to the loss of CO and 1,2,3,4-tetraphenylbenzene (TPB).

1.4 Synthesis of planar binuclear metallophthalocyanine dimer 10 [82a]

Planar or nearly planar binuclear Pc`s, binuclear Pc-triazolehemiporphyrazinates (Thp) [93] and phthalocyanine based dimers [93-96] are known. As an example, a dimer in which two Pc`s are linked with a bis(acetylene) bridge [94] was obtained by coupling the Pc derivatives having two acetylene units in the presence of a copper salt in pyridine. Another example is an oligo(phenylenevinylene)-bridged Pc dimer [96] obtained by our group through the reaction of 2 equivalents of a modified Pc-monoaldehyde with 1 equivalent of *p*-xylene bis(triphenylphosphonium)bromide. Also a Pc dimer fused with anthraquinone was prepared recently by our group.[95] Torres et al. synthesized a heterodimetallic binuclear Pc-derivatives, having Ni and Zn as different metals. [93] Preparation of heterodimetallic binuclear Pc-Thp-compounds with Ni and Zn have also been done by the same authors.[92]

The synthesis and spectroscopic characterization of the symmetrically substituted binuclear metallophthalocyanine **10** (Scheme 13) containing two methyl groups on the epoxybenzo unit are described here. Compared to an earlier work reported by us [80a] where the absence of methyl groups on the epoxybenzo unit prevented the complete aromatization of the target dimer **10**, we could successfully dehydrogenate the dimer **10** to the fully conjugated compound **11**.

Thermolysis of precursor **9** at 120 °C in xylene leads to the loss of CO and 1,2,3,4-tetraphenylbenzene with in situ generation of the reactive intermediates **9a** (Scheme 13). The DA reaction of the isobenzofuran intermediate **9a** (generated from **9**) with **4a** afforded the phthalocyanine dimer **10** (R = 2-ethylhexyl) in 25% yield (Scheme 13).

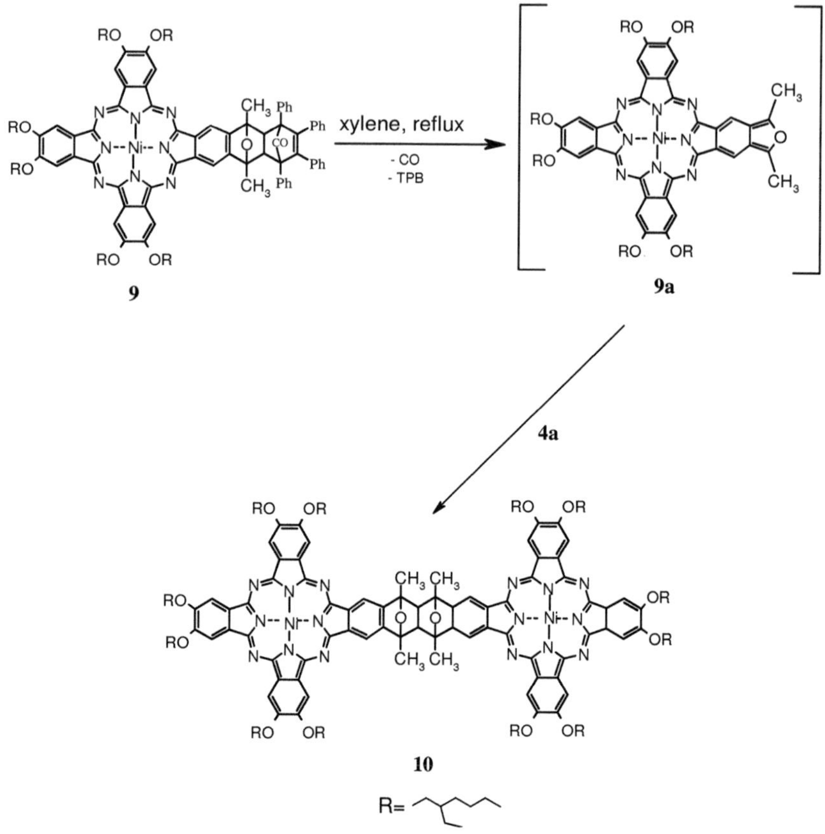

Scheme 13: Synthesis of binuclear metallophthalocyanine **10**

1.4.1 Spectroscopic characterization of phthalocyanine dimer 10

The ¹H-NMR spectrum of **10** exhibits relatively broad peaks of the Pc substituents (2-ethylhexyloxy) groups and broad aromatic signals in the region between 8.00 and 10.20 nm. The broadening is caused by the essentially flat nature of the molecules leading to aggregation. Compound **10** with its symmetric structure. The protons from C-2 and C-7 carbons appear at 8.82 ppm (see Figure 20 for numbering), while the C-10 protons are assigned at 8.96 ppm. The C-15 protons are detected at 10.30 ppm.

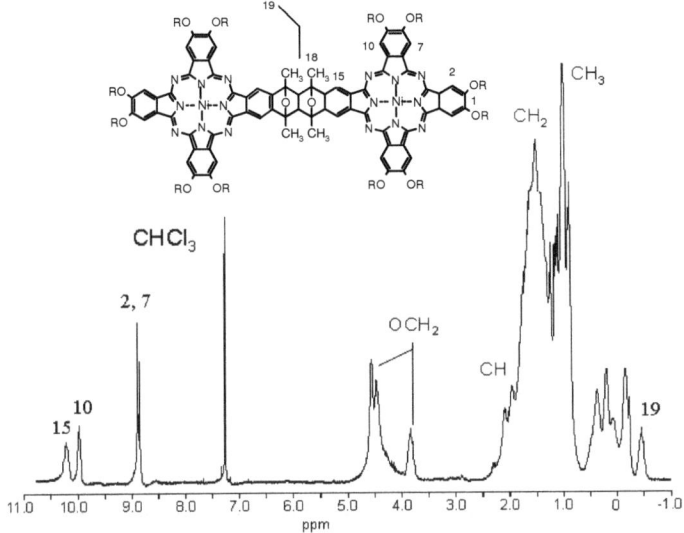

Figure 20: ¹H-NMR spectrum of phthalocyanine dimer **10**

Also the ¹³C-NMR spectra of the same type of compounds are quite different. The symmetry present in structure of **10**, leads to less similar peaks, the aromatic carbon peaks for C-2, C-7 and C-10 appear around 126 ppm. Carbons C-17 are allocated at 80 ppm, while the C-3, C-6, C-11, C-14 and C-16 carbons are assigned between 127.0 and 130.5 ppm. The C-4, C-5, C-12 and C-13 carbons are detected between 133.0 and 140.0 (Figure 21).

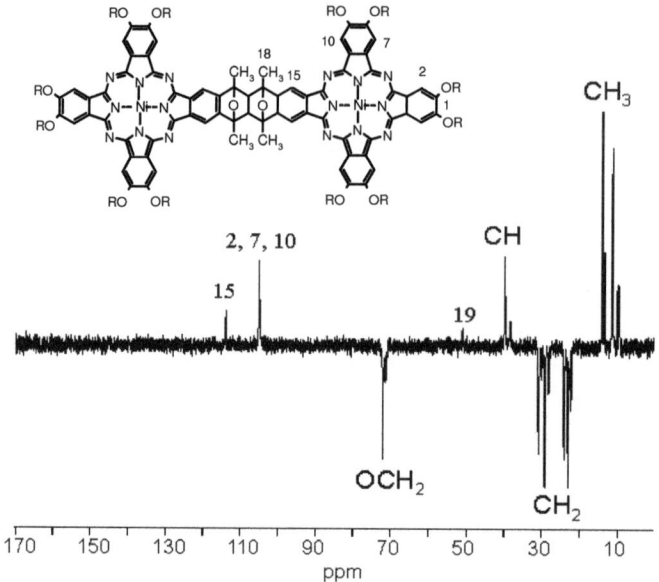

Figure 21: ^{13}C-NMR spectrum of **10**

The UV/Vis spectrum of **10** in CH$_2$Cl$_2$ shows the Q-band maxima at 668.5 nm, (Figure 22). In comparison to the monomeric (RO)$_8$PcNi (**AAAA** product, R = 2-ethylhexyl), broad absorptions in the Q-band region can be seen in **10** due to aggregation, which differ characteristically from the sharp peaks of the monomeric phthalocyanine **3a**. No or only a small red shift is observed for the Q-bands of **10**. This points to little π-electron delocalisation in the binuclear systems **10**, showing an almost independent behaviour of the two Pc-rings, in terms of their UV/Vis spectra. When comparing the spectra of **10** and (RO)$_8$PcNi (**3a**), a blue shift from the (RO)$_8$PcNi (670.0 nm) to **10** (668.5 nm) is seen.

1.5 Synthesis of dehydrated planar metal phthalocyanine dimer 11

By treating dimer **10** with a 6-fold excess of *p*-toluenesulfonic acid in toluene at 80 °C the fully unsaturated dimer **11** (R = 2-ethylhexyl) was obtained in a yield of 22%, after purification by flash chromatography. During the reaction a monodehydrated dimer was also observed on TLC control (Scheme 14).

Scheme 14: Synthesis of dehydrated Pc dimer **11**

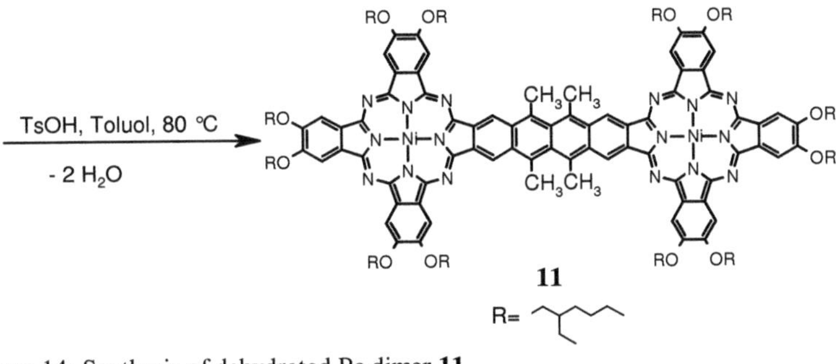

1.5.1 Spectroscopic characterization of dimer phthalocyanine 11

An analytically pure sample of **11** was identified by Maldi-Tof MS. It showed a molecular mass of m/z = 2809.16 corresponding to its structure. The UV/Vis spectrum of the dehydrated dimer **11** (in CH_2Cl_2) shows only a small influence in comparison to the nondehydrated dimer **10** on the wavelength values of the electronic transitions. The Q band of dimer **10** appears at 668 nm, for the corresponding dehydrated dimer **11** it was found at 690 nm (Figure 22). This point to the fact that, the two Pc rings in **11** are electronically separated from each other.

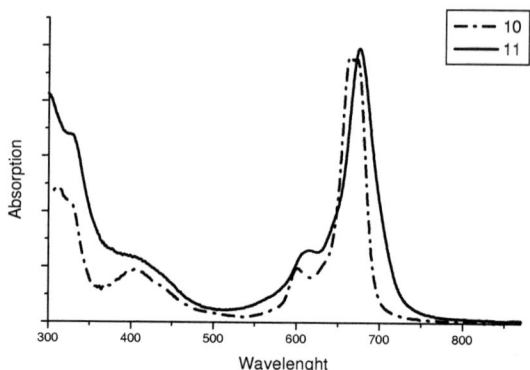

Figure 22: UV/Vis spectra of compounds **10** (----) and **11** (——) for comparison

As described above, the presence of the methyl groups on the epoxide ring in dimer **10** prevent the formation of hydroxyl groups of the target molecule **11** as previously described. [80]

2. Synthesis of planar trinuclear metal free phthalocyanine 13

We have reported the preparation of ladder-type oligomers earlier from Diels-Alder monomers based on Hps or Pc`s systems that can be used as bisdienophiles (Figure 6b, page 12).[80, 97] The formation of ladder-type oligomers occurs by generation of intermediates and reaction of macrocyclic bisdienophiles, these trimers contain three linked Pc macrocyclic units, which are isomeric mixtures.

Among the reported trinuclear Pc`s, the metal free trinuclear compound **13** with twelve (2-ethylhexyloxy) substituents was synthesized (Scheme 16).

The complicated synthetic pathway starts from the bis-dienophillic metal free Pc **5b** with ABAB symmetry. (see section 2.1).

2.1 Synthesis of symmetrically substituted bis-dienophillic metal free phthalocyanine 5b with ABAB-symmetry:

In general, as previously discussed (pages 27-29), in a reaction of two different dinitriles **A** and **B**, six different products can be expected. But for the synthesis of the trinuclear **13** (Scheme 16) by using the repetitive Diels-Alder reaction (DA) via the isobenzofuran route [97] which we have described earlier, [80, 81] the necessary starting material to carry out a straightforward synthesis of target trimer **13** is the symmetrically substituted bisdienophilic phthalocyanine PcH$_2$ **5b** (see Scheme 11). By changing the ratio between the two dinitriles **1** and **2** in the statistical synthesis into **1(A): 2(B)** = 1:1, the resulting amount of each isomer will be varied, so that the required isomer **5b** is obtained in good yield in comparison with previous cases.

The reaction of phthalonitriles **1** and **2** results in the formation of six phthalocyanines, viz. **AAAA (3b)** (self-condensation of **1**), **AAAB (4b), ABAB (5b), AABB (6b), ABBB (7b)** and **BBBB (8b)** (self-condensation of **2**) (Scheme 11). The main products isolated were **5b** and **6b**; The blue green compound **5b** was separated from the rest of products by flash chromatography starting with DCM–hexane (3:1) and finally DCM–ethyl acetate (2:1) as the mobile phase (see experimental part). PcH$_2$ **5b** was obtained in yield of up to 17 %.

2.1.1 Spectroscopic characterization of phthalocyanine 5b

The ^{1}H-NMR spectrum of **5b** exhibits two singlets for the protons of the epoxybenzo units 2-H and 1-H at δ = 6.2 and 7.30-7.38, respectively, in addition to the sharp singlets of the NH- at δ = -2.4, -2.5 (see Experimental part). Aggregation of **5b** in solution leads to broad resonances in its ^{1}H-NMR spectra, in addition to the signals of the alkyloxy chains in the nonaromatic region between 0.95 and 2.10 ppm,

The symmetry present in the structure of **5b**, leads to somewhat similar peaks in the ^{13}C-NMR spectrum in spite of its aggregation, showing the aromatic carbon peaks for C-9 and C-4 around 104 and 113 ppm, respectively. Carbons C-8 and C-6 are allocated at 130 and 149.1 respectively, while the alkyloxy substituted aromatic carbons C-10 is assigned around 152 ppm.

2.2 Synthesis and spectroscopic characterization of Pc-tetracyclone bisadduct 12

Compound **5b** and excess tetraphenylcyclopentadien-1-one (tetracyclone) were dissolved in dry toluene and stirred at 70 °C for 3-4 days. Compound **12** was obtained in 84% yield after chromatographic work-up.

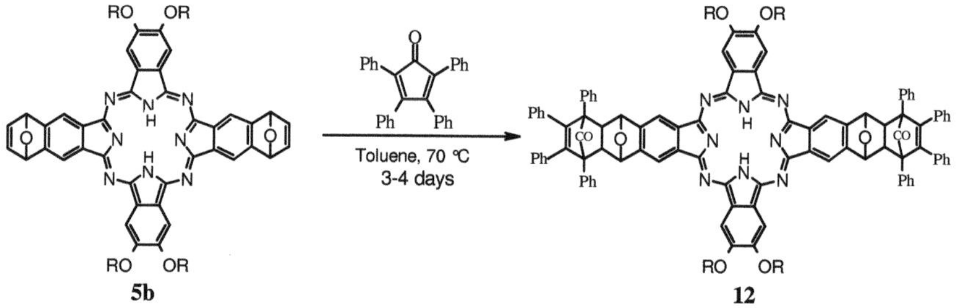

Scheme 15: Synthesis of Pc-tetracyclone bisadduct **12**

The ^{1}H-NMR spectrum of **12** shows the characteristic signals of the phenyl protons in the aromatic region, between 7.10 and 7.67 ppm, as well as the appearance of singlets for 3-H in **12** at δ = 3.4 which is in agreement with the expected *exo*-orientation of the tetracyclone adduct Additional signals for the alkyloxy chains are in the nonaromatic region, between 0.91 and 2.30 ppm. while the ^{13}C-NMR spectrum presents the characteristic C=O signal at 196.6

ppm, among the other assigned signals (see experimental, page 83).

The UV/Vis spectrum shows a five nm red-shift, from 690.0 in **5b** to 695.0 nm in **12**.

The IR spectrum of **12** also reveals clearly the C=O band at 1772 cm^{-1}.

The MALDI-TOF MS of **12** shows two fragments peaks at m/z = 1928.46 (100) M$^+$.

2.3 Synthesis of trinuclear metallophthalocyanine trimer 13 [82b]

For the synthesis of the trimer **13,** we used a repetitive Diels-Alder reaction (DA) via the isobenzofuran route as we have described earlier.[81, 82b] starting with the bisdienophilic Pc **5b**. Thermolysis of **12** at 120°C in toluene leads to loss of CO and 1,2,3,4-tetraphenylbenzene (TBP) with in situ generation of the reactive intermediate **12a**, which reacts with a 3 fold excess of the bisdienophile **5b** to form the trimer **13** in 25% yield (Scheme 16).

Trimer **13** is a mixture of isomers concerning the position of the oxygen bridges. We did not attempt to separate the isomers, according to earlier results the separation of these kind of isomers is very difficult.[82b] Purification of the trimer **13** was performed, by extracting several times with methanol and acetone to remove TPB and excess of **5b.**

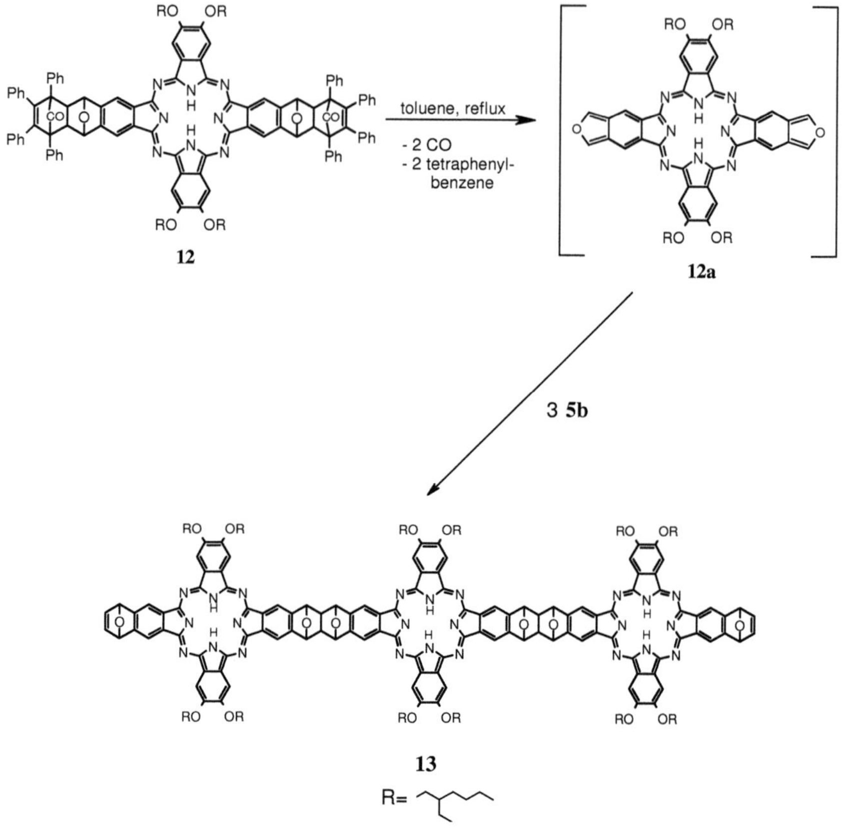

Scheme 16: Synthesis of trinuclear phthalocyanine **13**

2.3.1 Spectroscopic characterization of 13

The ¹H-NMR spectrum of phthalocyanine **13** shows, as pointed out before for the corresponding Ni-dimer **10**, the predictable aggregation and broadening of the signals. The identification and assignment of the peaks became easier when the ¹H- and ¹³C-NMR spectra were recorded in deuterated THF as solvent. In the ¹H-NMR spectrum, the 2-ethylhexyloxy substituents appear between 0.99 and 2.15 ppm, plus the OCH$_2$ groups at 5.00 ppm. Characteristic resonance of δ = 3.5 (H-20) (the linkage between the macrocycles), and the other signals are in agreement with the ¹H-NMR data of the macrocyclic monomer **5b** (see experimental, page 86).

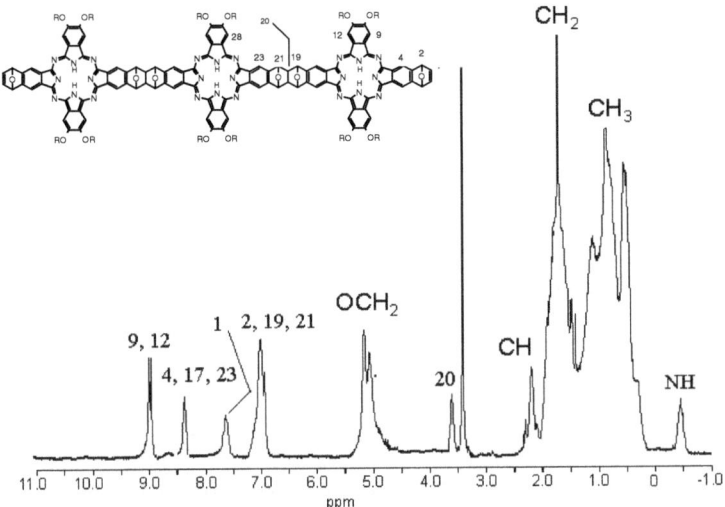

Figure 23: ¹H-NMR spectrum of phthalocyanine trimer **13**

The ¹³C-NMR spectra of **13** shows extremely broad and weak resonances, due to aggregation of the molecules, The ¹³C-NMR spectrum is also very similar for the compound **10**. The assigned carbon peaks appear in the same region for these compounds (see experimental, page 86). It shows the aromatic carbon peaks for C-9, C-12 and C-17 in the region around 105 ppm. The C-2 and C-19 carbon atoms of the epoxy rings can be assigned around 79 ppm.

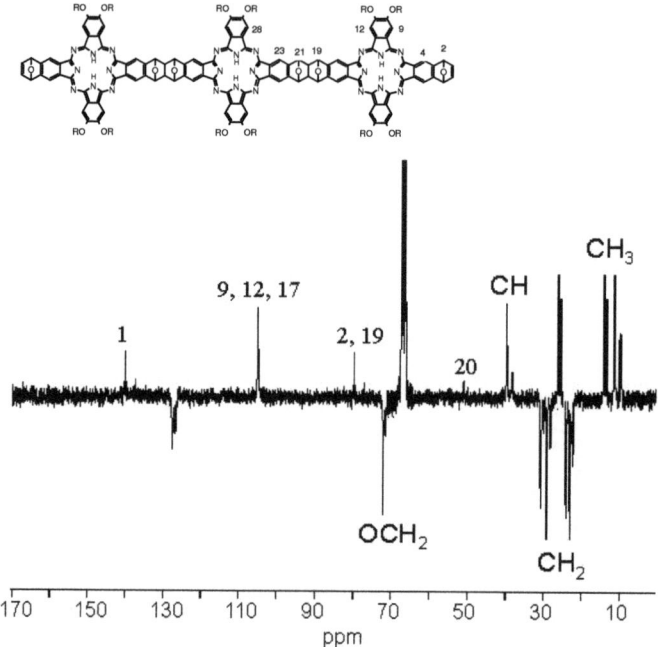

Figure 24: ^{13}C-NMR spectrum of trimer **13**

The UV/Vis measured in CH_2Cl_2 shows two Q-band maxima at 699 and 661 nm. Broad absorptions in the Q-band region are observed for **13** due to aggregation.

3 Synthesis of pyridine *N*-oxide adducts 15a-c.[98]

Reactions of asymmetrically substituted dienophilic hemiporphyrazines and phthalocyanines e.g. **4a** and similar Pc`s [99, 100] with dienes to provide, e.g. materials for the preparation of conductive phthalocyanine polymers or compounds for nonlinear optical applications have been studied in our group quite intensively.[83,84,97,99]

4a as pointed above was especially used in Diels-Alder reactions e.g. with tetraphenylcyclopentdienone to afford the Pc-tetracyclone adduct, which is a precursor to synthesize binuclear or trinuclear Pc`s.[82]

To study additional cycloaddition reactions of **4a**,[82] we studied here its reaction with the pyridine *N*-oxides **14a-c**. Through this reaction a series of phthalocyanines **15a-c** containing fused furopyridine rings are obtained (Scheme 17). Furopyridines are targets for pharmacological studies, e.g. for treatment of several infectious diseases. [101]

The reaction of electron-deficient pyridine *N*-oxides in inverse-type 1,3-cycloaddition reactions with dipolarphiles e.g. functionalized olefins, cumulenes, acetylenes,[102] phenyl isocyanates, [103] and *N*-phenylmaleimides,[104] have been investigated in detail by Hisano and co-workers.[105] The reaction of pyridine *N*-oxides with 1,4-epoxy-1,4-dihydronaphthalene to afford the aromatized furopyridine-type cycloadducts was also investigated by the same authors. [105]

General synthesis

PcNi **4a** with nickel as the central metal ion and (2-ethylhexyloxy) as substituents were chosen in order to facilitate chelating, to achieve good solubility and to suppress aggregation. 3-Methylpyridine-*N*-oxide (**14a**), 3-carboxypyridine-*N*-oxide (**14b**) and 4-cyanopyridine-*N*-oxide (**14c**) were reacted with PcNi **4a** forming the adducts **15a-c** (Scheme17) in 52-75% yield (see below).

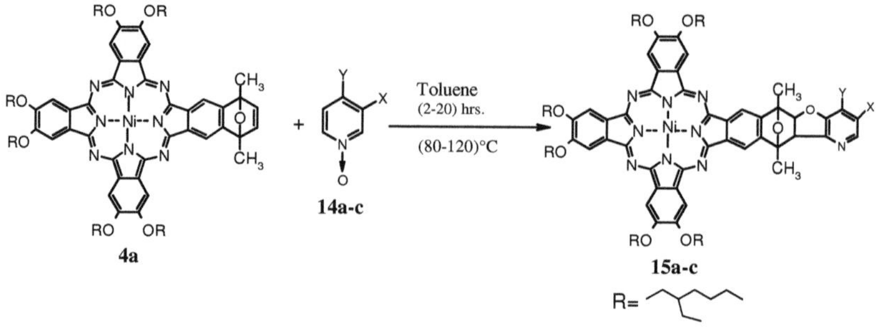

4a

14a-c

15a-c

Scheme 17: Synthesis of pyridine *N*-oxides **15a-c**

The nickel phthalocyanines **15a-c** are formed via 1,3-dipolar cycloaddition of the pyridine *N*-oxides **14a-c** with the 1,4-epoxy-1,4-dimethyl-1,4-dihydrobenzene unit in PcNi **4a** to give intermediate *endo-exo* cycloadducts. Thermal conversion of the intermediate adducts to the products **15a-c** takes place via a 1,5-sigmatropic rearrangement as shown in Figure 25.

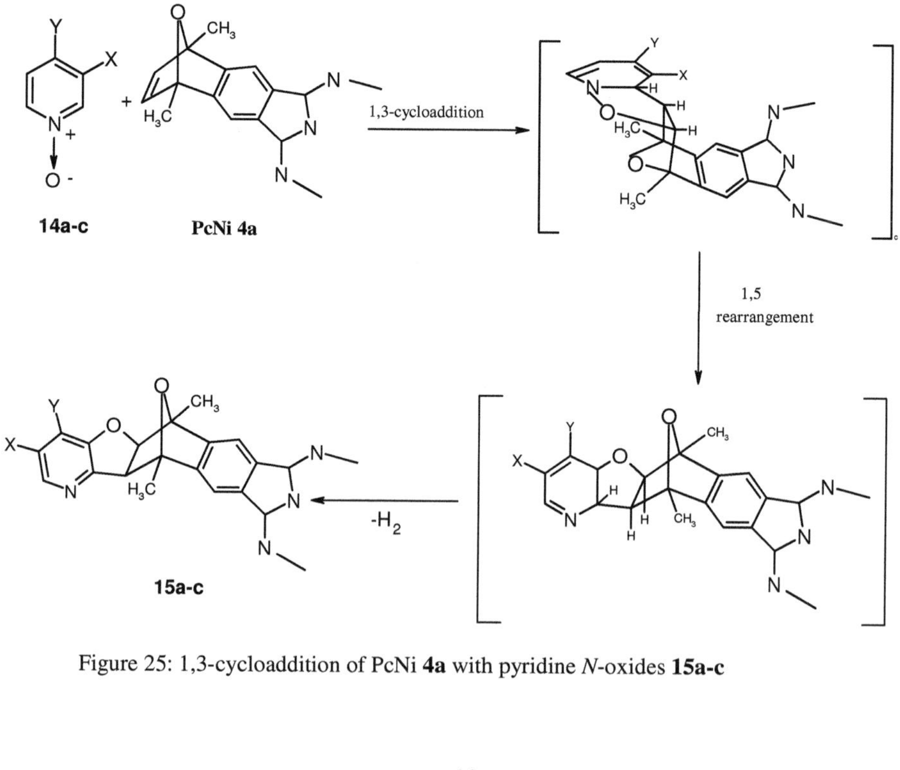

Figure 25: 1,3-cycloaddition of PcNi **4a** with pyridine *N*-oxides **15a-c**

The starting material PcNi **4a** and the pyridine *N*-oxides reagents were dissolved in 20 mL of the appropriate solvent and stirred at different temperatures (TLC control) for different times (Table 2). Isolation and purification of adducts **15a-c** were performed by flash chromatography (see experimental part). The reactions were performed in an autoclave always using an excess of **14a-c**.

Table 2: Reaction of pyridine *N*-oxides **14a-c** with PcNi **4a**

Pyridine *N*-Oxides			Reaction Conditions			Product			Yield (%)
	X	Y	Solvent	Temp.(°C)	Time	Adduct **15a-c**	X	Y	
14a	CH_3	H	Toluene	80	2	**15a**	CH_3	H	52
14b	COOH	H	Xylene	120	20	**15b**	H	H	58
14c	H	CN	Xylene	110	4	**15c**	H	CN	75

According to Table 2 only a moderate difference in reactivity in case of pyridine *N*-oxides **14a** and **14c** is obtained in spite of the fact that the yield of **15c** with somewhat higher reaction time is larger than the yield of **15a**. Pyridine *N*-oxide **14b** containing the COOH group shows a low reactivity towards the 1,4-epoxy-1,4-dimethyl-1,4-dihydrobenzene unit, which might be due to the stabilization of the 1,3-dipole by intermolecular hydrogen bonding on the oxygen atom of the N → O group.

3.1 Spectroscopic characterization of pyridine N-oxide adducts 15a-c

The purified products **15a-c** were identified by their NMR, IR and FD- or FAB-MS spectra (see Experimental part).

General characteristic signals in the ^{1}H NMR spectra for the furopyridine units in **15a-c** are 1-H/7-H at δ = 4.41 and 3.98 ppm and 3-H/4-H and 5-H of the pyridine ring.

In addition, strong and well resolved resonances are observed for the methyl groups on the epoxybenzo ring at δ = 1.2 for **15a-c**. The presence of the methyl groups increase the solubility and stability properties in both of PcNi **4a** and Pc`sNi **15a-c**. (see Experimental part)

For example, the ¹H NMR data and integral values for **15b** and the obtained data are consistent with it´s chemical structure (see Figure 26).

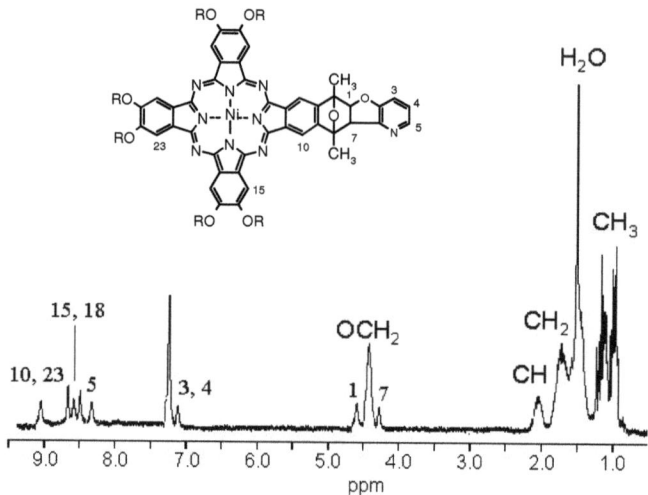

Figure 26: ¹H-NMR spectrum of **15b**

The UV/Vis spectra of the produced adducts **15a-c** are quite similar in shape in comparison with the starting material PcNi **4a**, except for small red shift of the Q-bands between 1 and 6 nm. The formation of the adducts **15a-c** shows that the reactions of **4a** described in the introduction can be extended also for the attachment of other heterocyclic rings to the phthalocyanine macrocycles.

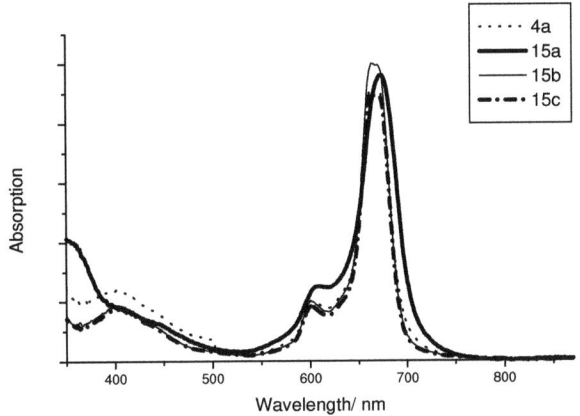

Figure 27: UV/Vis spectra of adducts **15a-c**

In summary, we have successfully demonstrated the usage of an effective building block for synthesizing a series of geometrically variable nickel phthalocyanine complexes using pyridine *N*-oxides. The described route is easy and all the products obtained were purified and well characterized. From our results, the use of the Diels-Alder strategy as a novel and general approach to asymmetrically substituted monoadducts of nickel phthalocyanine systems **15a-c** has proved to be efficient and convenient.

4 Synthesis and properties of phthalocyaninatoindium(III)acetylacetonates 16, 19 and 20 designed for optical limiting purposes [82b]

Pc`s are especially attractive because of their NLO properties which can be modified via suitable structural modifications (pages 20-23 ff). The change of the central atom (metal) in phthalocyanines can lead to a considerable variation of the relevant NLO properties for OL. Another significant factor influencing the OL properties of Pc`s is the variation of the axial substituents in phthalocyanines (see pages 21-23 ff).

Most of the axial substituents which have been used so far for this purpose are e.g. halogen (mostly Cl), p-CF$_3$ phenyl, oxygen e.g. in PcTiO and derivatives theirfrom. [106-109]

For the first time we introduce here a new axial substituent in metal phthalocyanines, namely acetylacetonate (acac) with In^{3+} as central atom e.g. in **16** and **19** and trimer **20** to study the NLO properties. [82b]

These compounds will be compared with other described axially pc`s with chloro, aryl substituted In(III)phthalocyanines concerning their OL properties.

The parent compound of indium phthalocyanines, the unsubstituted PcInCl, has been prepared starting from phthalonitrile and indium(III)chloride in high-boiling solvents like quinoline,[110] or 1-chloronaphthalene,[111] or from PcLi$_2$ and indium(III)chloride in 1-chloronaphthalene.[112]

The greatest disadvantage of peripherally unsubstituted phthalocyanines is their poor solubility in common organic solvents. To overcome this problem, a variety of substituents have been attached to the macrocycle, in varying numbers and different substitution patterns (see page 22 ff).

To obtain the substituted phthalocyanines **16**, **19** and **20** (Schemes 18-20) with In(acac) as central moiety is expected to be difficult, due to the poor stability of this moiety against the number of steps which are necessary to obtain also the corresponding trinuclear metal phthalocyanines (see page 54).

To solve this problem an approach was used in which Inacac is inserted only in the final step, avoiding the expectable big losses of material throughout all steps shown in Schemes 18-20 (pages 51-54).

4.1. Synthesis of Octa-(2-ethylhexyloxy)phthalocyaninatoindium(III) acetylacetonate (16)

For the first time we synthesized an octaalkoxy substituted indium Pc compound with acac as axial substituent, namely $(RO)_8PcInacac$ with R = 2-ethylhexyl.

$(RO)_8PcInacac$ **16** (R = 2-ethylhexyl) was prepared by reacting $(RO)_8PcH_2$ **3b** (R = 2-ethylhexyl) with $In(acac)_3$ in DMF at 140°C (Scheme 18). Compound **3b** was collected as the first fraction from the chromatographic separation of the products shown in Scheme 11 (page 29)

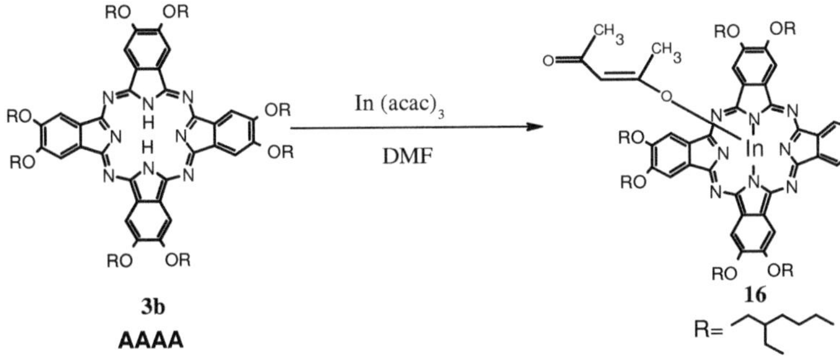

Scheme 18: Synthesis of $(RO)_8PcInacac$ **16**

4.1.1 Spectroscopic characterization of 16

In the ^{1}H-NMR spectrum, the 2-ethylhexyloxy substituents appear between 0.85 and 2.05 ppm, plus the OCH_2 groups at 4.15 ppm. No NH-protons could be seen in the ^{1}H NMR spectrum of a pure sample of Pc **16**.

In the IR spectrum of **16** the absorption of the carbonyl band appears at 1726 cm^{-1} (see Experimental part). The split Q-band in the UV/Vis spectra of a dilute sample of $(RO)_8PcInH_2$ **3b** (toluene) is centered at 662 and 700 nm, the single Q-band of $(RO)_8PcInacac$ **16** appears at 696 nm.

4.2 Synthesis of tetra-*tert*-butylphthalocyaninatoindium(III)acetylacetonate (19)

The method used for the preparation of the tetra-*tert*-butylphthalocyaninatoindium(III) acetylacetonate **19** was similar to the one used for the preparation of **16**. Tetra-*tert*-butylphthalocyaninatoindium(III)acetylacetonate, (*t*Bu)$_4$PcInacac **19** was obtained in good yield by reacting the metal-free (*t*Bu)$_4$PcH$_2$ **18** [113] with an excess of In(acac)$_3$ in refluxing DMF (Scheme 19).

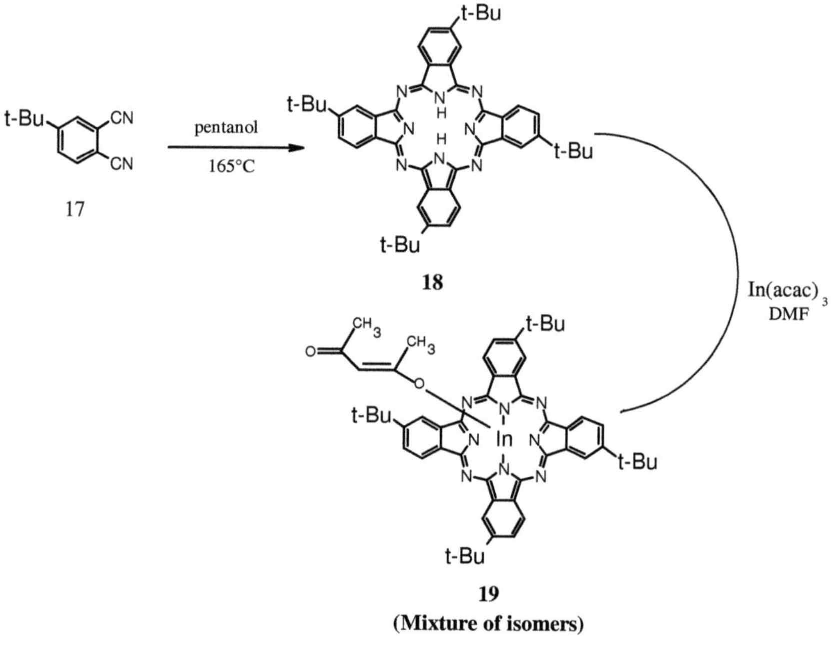

Scheme 19: Synthesis of (*t*Bu)$_4$PcInacac **19**

4.2.1 Spectroscopic characterization of 19

The ^{1}H-NMR spectrum reveals that the chemical shifts of the protons of the macrocycle for **19** are close to the corresponding values for the respective chloro complex (*t*Bu)$_4$PcInCl **21a**. This shows that there is only little influence of the axial ligand (acac) on the electronic structure of the complex **19**.

19 exhibits three multiplets in the aromatic region at δ = 8.38, 9.41 and 9.46 for the 1`, 2 and the 2` protons, respectively (see Experimental part).

The *tert*-butyl groups of **19** give rise to absorptions in the IR-spectrum due to the C-H stretching mode in the range 2867-2958 cm^{-1} and the C-H deformation modes in the 1486-1330 cm^{-1} range. Comparing the IR spectrum of **21a** and **19** reveals that they are almost identical. The only difference is the appearance of the In-L vibration. In the case of the axially chloro substituted complex, a weak absorption band of the In-Cl stretching mode appears at 336 cm^{-1}, in case of **19**, the corresponding C=O-absorption stretching frequencies are shifted to 1726 cm^{-1}.

The UV/Vis absorption spectrum of **19** reveals that the influence of the axial (acac) on the electronic structure of the macrocycle is only small. The UV/Vis spectrum of a dilute sample of metal-free (*t*Bu)$_4$PcInH$_2$ **18** (recorded in toluene) shows the splitted Q-bands centered at 663 and 700 nm [113] For (*t*Bu)$_4$PcInacac **19**, the Q-band appears without shifting at 671 nm. (Figure 28). The axial (acac) ligand has little influence on the position of the non splitted Q-band maxima in comparison with the chlorine ligand in **21a** [114] (697 nm) (Figure 28).

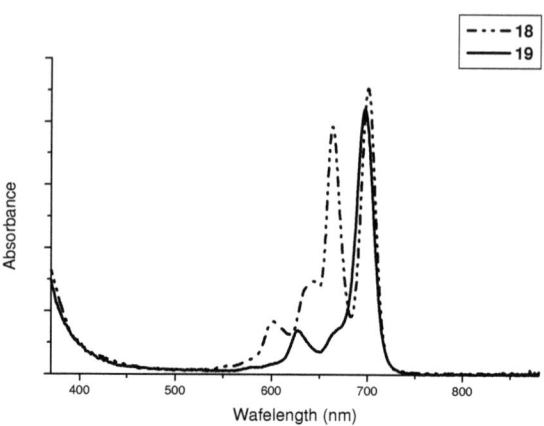

Figure 28: UV/Vis spectra of compounds **18** (----) and **19** (——) for comparison

4.3 Synthesis of trimeric phthalocyaninatoindium(III)acetylacetonate 20

Trimer **20** was obtained by reacting the trimeric metal-free PcH$_2$ **13** with an excess of In(acac)$_3$ in refluxing DMF. After completion of the reaction (monitored by UV/Vis spectroscopy and thin-layer chromatography), water was added dropwise to the mixture to precipitate the compound. After further purification (see Experimental part), trimer **20** was obtained as a blue green powder.(Scheme 20).

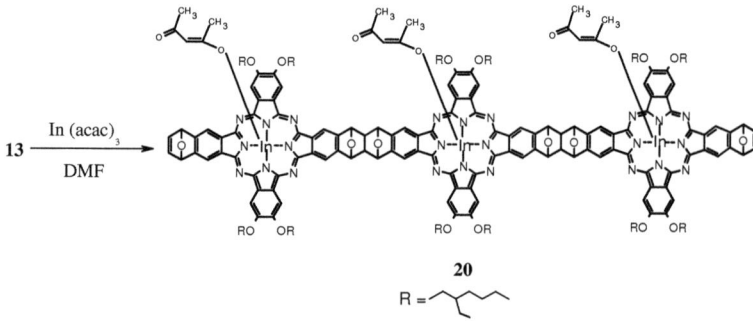

Scheme 20: Synthesis of trimeric PcInacac **20**

4.3.1 Spectroscopic characterization of 20

The solubility of **20** in organic solvents is comparatively low, the structure of **20** was established by ^{13}C CP/MAS spectroscopy (Figure 29).

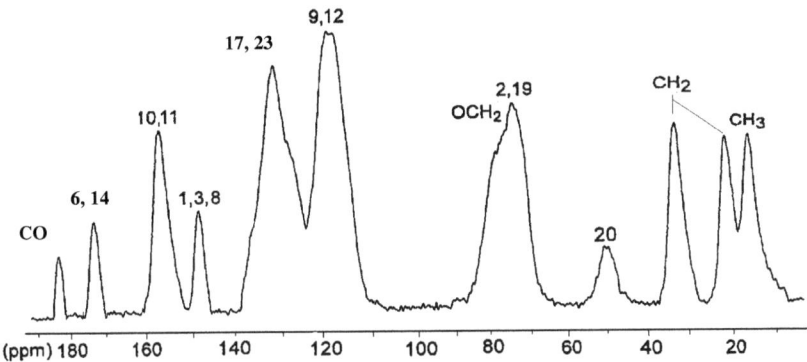

Figure 29: ^{13}C CP/MAS spectrum of phthalocyanine trimer **20**

The characteristic signal of C-20 is found at $\delta = 51.00$, the other signals are in agreement with the ^{13}C NMR data of the macrocyclic monomer **5b** (see Experimental part).

The UV-Vis spectrum of **13** in toluene shows the splitted Q-band centered at 661 and 699 nm, in **20** the nonsplitted Q-band is observed at 686 nm (Figure 30).

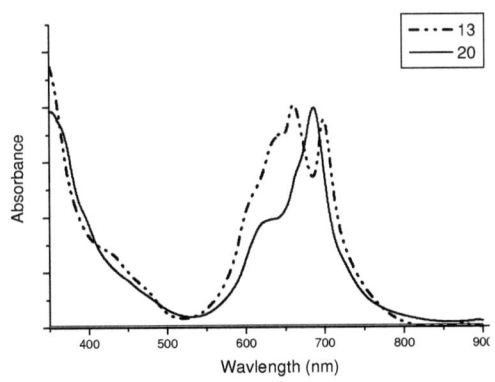

Figure 30: UV/Vis spectra of compounds **13** (----) and **20** (——) for comparison

Table 3 shows the Q and B-band absorptions of the newly synthesized compounds for comparison.

Table 3: Q and B band absorptions for the list of compounds.

Compound	Q(nm)		B(nm)	
16	696	627	404	350
19	671	607	410	361
20	686	605	420	350
21a	696	665	359	339

* All spectra were recorded in CH_2Cl_2 as solvent

From the spectra of the trimer **20** nothing can be concluded about the relative position of the (acac) substituent to each other, syn or anti. Also from the chromatographic results no indication was obtained about the relative compositon of the possible isomers.

5 Nonlinear optical properties of phthalocyaninatoindium(III)acetylacetonates 16, 19 and 20 [82b]

As described on pages 50-56. Pc`s have been extensively investigated recently for their nonlinear optical (NLO) properties of which optical limiting (OL) is closest to practical applications [115-120]

We have synthesized earlier the soluble axially substituted arylphthalocyaninatoindium complexes, listed in Table 4 and studied the effect of the axial substituent L on the optical limiting behaviour.[121, 124] Pc`s **21b-e** were found to be better optical limiters than the chloro compound **21a**.[121] Axial substitution alters the electronic structure of the Pc through the presence of an additional dipole moment oriented perpendicularly with respect to the macrocycle.[125] Axial substitution also diminishes aggregation in solution. Both effects strongly influence the OL properties of these compounds.[120,121,126]

We also prepared the corresponding gallium phthalocyanines (tBu)$_4$PcGaL (L = Cl, p-TMP) and μ-oxo-bridged Pc-gallium compounds [R$_n$PcGa]$_2$O [126] and investigated their OL properties.

Table 4: Chloro- and arylindium(III)phthalocyanines

21	L	abbreviation
a	- Cl	
b	-C$_6$H$_4$-CF$_3$ (CF$_3$)	p-TMP
c	-C$_6$H$_4$-CF$_3$	m-TMP
d	-C$_6$H$_4$-F	p-FP
e	- C$_6$F$_5$	PFP
f	- Ph	

In the following, the optical limiting properties of octa-(2-ethylhexyloxy)phthalocyaninatoindium(III)acetylacetonate (**16**), tetra-*tert*-butylphthalocyaninatoindium(III) acetylacetonate (**19**) and trimer **20** are compared with the ``state of art`` molecule, tetra-*tert*-butylphthalocyaninatoindium(III)chloride (**21a**).[120, 123]

5.1 Z-Scan measurements

The measurement techniques used in general to study NLO-effects are third-harmonic generation (THG),[127] degenerate four-wave mixing (DFWM),[128-131] electric field induced second harmonic generation (EFISH),[132] and Z-scan methods.[133,134]

The Z-scan method will be discussed in the following in more detail, in connection with themeasurements of the OL effect of the compounds **16**, **19** and **20**.

The Z-scan technique [133, 134] allows the experimental determination of the nonlinear transmission and nonlinear refraction. In the Z-scan technique the sample under investigation moves along the optical axis of a focused Gaussian beam. The sample experiences a large variation of the incident intensity along its path and nonlinear optical effects can be then induced.

Considering a Gaussian beam in a tight focus geometry as shown in Figure 31, the transmittance of a nonlinear medium is recorded through an open or closed aperture as a function of the sample position z measured with respect to the focal plane (z = 0). The sample for a Z-scan determination must have a thickness smaller than the diffraction length w_0 π / λ (w_0 is the beam waist radius at the focus and λ is the laser wavelength) of the focused beam (thin medium condition).

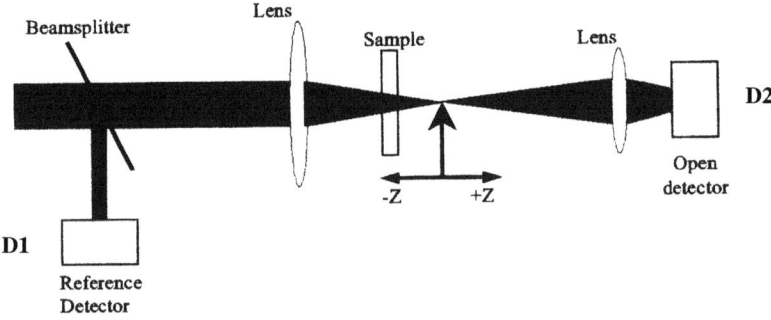

Figure 31: Typical Z-scan set-up. The ratio of the signal measured by the photo diodes D2/D1 is recorded as a function of sample position z

The Z-scan is completed as the sample is moved away from focus (positive z) such that the transmittance becomes linear since the irradiance is again low. Therefore, a prefocal transmittance minimum (valley) followed by a post focal transmittance maximum (peak) is the Z-scan signature of the sole positive refractive nonlinearity as shown in Figure 32 when a Z-scan closed aperture configuration is adopted (presence of a spatial filter in front of D2 withan aperture smaller than the beam diameter in Figure 31). Negative nonlinear refraction, following the same analogy, gives rise to an opposite peak-valley pattern. The sign of the nonlinear index is then immediately obvious from these patterns (Figure 32).

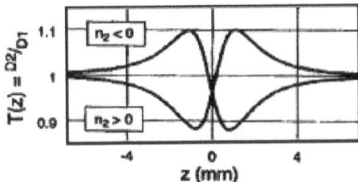

Figure 32: Z-scan signature for the sole occurrence of negative (positive) refractionnonlinearity resulting from a prefocal beam narrowing (broadening) followed by a postfocal beam broadening (narrowing).

In case of positive nonlinear absorption, i.e. occurrence of RSA, then in the open aperture configuration (presence of a spatial filter in front of D2 with an aperture larger than the beam diameter in Figure 31), the typical Z-scan pattern is that presented in Figure 33. As the sample approaches the beam focus the increase of beam intensity promotes changes in the sample in such a way that the overall absorption coefficient of the sample increases and, consequently, the transmission decreases. In the far field region, i.e. $Z \gg 0$ or $Z \ll 0$, which corresponds to the linear optical regime.

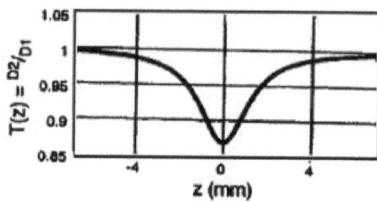

Figure 33: Z-scan pattern of a positive nonlinear absorber (RSA occurrence).

The open aperture of a Z-scan experiment [135] was used to measure the optical limiting response in the samples. All experiments described in this study were performed using 6 ns 532 nm laser light pulses from a Q switched frequency doubled Nd:YAG laser with a pulse repetition rate of 10 Hz. The beam was spatially filtered to remove the higher order modes and tightly focused using a 9 cm focal length lens. All samples were measured in quartz cells with a 1 mm optical path length, and at concentrations of 0.5 g/L in spectroscopic grade toluene.

5.2 Optical limiting measurements of 16, 19 and 20

All Z-scans performed exhibit a decrease of transmittance about the focus typical of an induced positive nonlinear absorption of incident light. The nonlinear absorption coefficient, β_I, experienced by the incident pulses was determined from these spectra as a function of the on focus intensity. Recently we have adopted this approach in determining effective values of β_I of gallium and indium phthalocyanines.[136, 137] The values of β_I for compounds **21a, 16, 19** and **20** as a function of the on focus intensity (I_0) are plotted in Figure 34. It can clearly be seen that as I_0 increases β_I tends to decreases in magnitude for all samples except for compound, tetra-*tert*-butylphthalocyaninatoindium(III)chloride (**21a**) which was measured over a larger intensity range.[136] At intensities beyond those in the figure its β_I coefficient also decreases with increasing intensity. It can be seen that the magnitude of the nonlinear absorption coefficient is largest, over the intensity region presented in the plot, for the compounds in order of **16 > 21a > 19 > 20**. The nonlinear absorption coefficient in all cases is of the order of $10^{-8} - 10^{-7}$ cm W^{-1}.

The optical limiting data plotted with the normalised transmission (T_{Norm}) against the incident energy density per pulse (J cm^{-2}) are depicted in Figure 35. The nonlinear absorption coefficient $\alpha(F, F_{Sat}, \kappa)$ where $\alpha(F, F_{Sat}, \kappa) = \alpha_0(1 + F/F_{Sat})^{-1}(1 + \kappa F/F_{Sat})$ derived from laser rate equations in the static state was used to fit the normalised transmission as a function of this energy density to a superposition of all open aperture datasets for each compound. In this expression F represents the energy density, F_{Sat} the saturation energy density and κ the ratio of the excited to ground state absorption cross sections σ_{ex}/σ_0. The parameters κ (realistically σ_{ex} as α_0 was measured) and F_{Sat} were treated as free constants in the fitting algorithm. The plots of normalised transmission against pulse energy density for compounds **21a, 19, 16** and **20** where the solid lines are theoretical fits to the experimental data are shown in Figure 35.

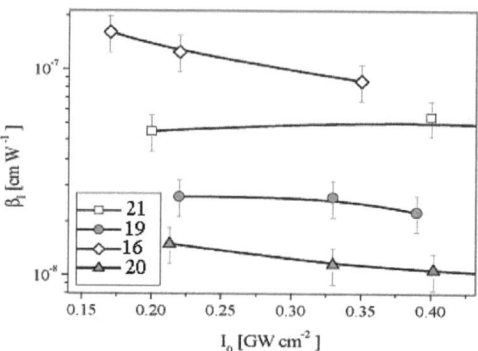

Figure 34: Plots of effective β_I against the on focus beam intensity I_0 for compounds **21a**, **16**, **19** and **20**. Each data point represents an independent open aperture z-scan, the solid lines are intended as guides to the eye.

The plot is expanded in Figure 35 about the section where the data almost overlaps for clarity. The α_0, κ and F_{Sat} values for each compound are presented in Table 5.

The F_{Sat} value for the axially substituted acetylacetonate indium monomer **19** is reduced significantly compared to that of compound **21a**. The addition of the acetyl acetonate group reduced the magnitude of the saturation energy density by a factor of approximately 2.1. This effect was far more dramatic through the modification of the $(tBu)_4PcInCl$ **21a**. The magnitude of the saturation energy density was reduced by more than one order of magnitude when **21a** was modified to **16**. Compound **20** has a far less dramatic effect reducing the F_{Sat} value by a factor of about 1.2 over compound **21a**.

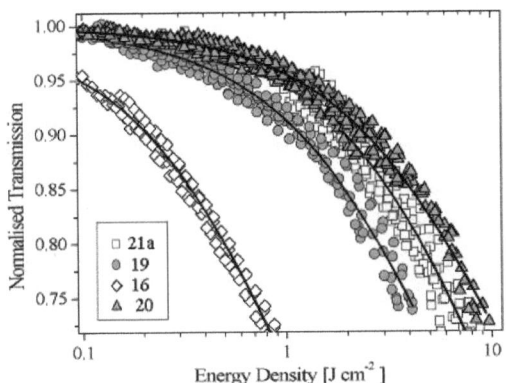

Figure 35: Plot of normalised transmission against pulse energy density for compounds **21a**, **16**, **19** and **20** at 0.5 g/L in toluene where the solid lines are theoretical fits to the experimental data. The fitting parameters are given in Table 5.

The ratio of the excited to ground state absorption cross sections $\kappa = \sigma_{ex}/\sigma_0$ was reduced from that exhibited by **21a** to each of the modified species presented here. The κ coefficient was reduced by a factor of 2.5 for modifications to **21a** producing **20** and by a factor of approximately 3.1 for **19**.

Table 5: Summary of the nonlinear optical properties for compounds **21a**, **16**, **19** and **20**. All measurements were performed with the compound dissolved in spectroscopic grade toluene at 532 nm.

Sample	Conc. [g L^{-1}]	α_0 [cm^{-1}]	κ [σ_{ex}/σ_0]	F_{Sat} [J cm^{-2}]
21a	0.5	0.53	27.4 ± 0.6	24.2 ± 0.8
19	0.5	1.5	8.7 ± 0.7	11.6 ± 1.4
16	0.5	1.3	10.8 ± 0.3	2.3 ± 0.1
20	0.5	0.99	11.0 ± 0.6	20.6 ± 1.5

6 Synthesis of binuclear ruthenium phthalocyanine [138]

The synthesis of pure (phthalocyaninato)ruthenium(II) (PcRu) by thermal decomposition of PcRu(DMSO)$_2$·2DMSO was reported by us already 17 years ago.[139] Later we described a more convenient method for the preparation of pure PcRu via the easy available bis-isoquinoline complex PcRu(iqnl)$_2$, [140] which can be thermally decomposed at 250°C. Further, tetra-*tert*-butylphthalocyaninatoruthenium(II) (*t*Bu)$_4$PcRu [141] was obtained by thermal decomposition of (*t*Bu)$_4$PcRu(3-Clpy)$_2$ (Clpy=3-chloropyridine) as a mixture of four structural isomers, 2,3-naphthalocyaninatoruthenium(II) (2,3-NcRu) as well as tetra-*tert*-butyl-2,3-naphthalocyaninatoruthenium(II) (*t*Bu)$_4$2,3-NcRu [141a] were obtained by thermal decompostion of the complexes 2,3-NcRu(L)$_2$ L= (3-Clpy) and (2-ethylhexyl)amine.[142] The *tert*-butyl substituted phthalocyanine and naphthalocyanine complexes are soluble in common organic solvents, which is also the case for several octaalkyloxy-phthalocyaninatoruthenium compounds.[143, 144, 145] PcRu has a dimeric structure with a Ru-Ru double bond distance of 2.40 Å as shown for the first time by Ercolani et al. [146] using large-angle X-ray scattering (LAXS) and magnetic measurements. On the basis of magnetic measurements our group proposed a metal-metal bond between two Ru-atoms for (*t*Bu)$_4$2,3-NcRu.[142] The dimeric structure for PcRu [147] and (*t*Bu)$_4$2,3-PcRu [141b] was confirmed by EXAFS spectroscopy. Temperature-dependent measurements of the magnetic susceptibility of (*t*Bu)$_4$2,3-PcRu exhibit paramagnetic behavior with strong coupling. The magnetic moment increases from 0.63μ_B (T= 20K) to 1.68μ_B (T= 300K), i.e. it approximates asymptotically the spin-only value of one unpaired electron (1.73μ_B), a value that is similar to PcRu.[141a] (*t*Bu)$_4$PcRu therefore has only one spin per ruthenium(II) ion, although for a d^6 transition metal ion in a square-planar ligand field two unpaired electrons are expected. This implies that (*t*Bu)$_4$PcRu also exists as a dimeric structure with a ruthenium-ruthenium double bond.[148, 149]

In our continuing investigations about the influence of strongly electron withdrawing substituents in the peripheral positions of the phthalocyanine ring on the physical and structural properties of metal phthalocyanines,[150] we carry out investigations on hexadecafluoro(phthalocyaninato)ruthenium F$_{16}$PcRu in order to proof whether or not this compound also has a dimeric structure.

The strongly electronegative and π–electronwithdrawing fluorine substitutents in the peripheric positions of the $F_{16}PcRu$, in principle, could change the bonding between the two PcRu units.

The $F_{16}RuPc$ complex was prepared earlier by reaction of tetrafluorophthalonitrile and $Ru_3(CO)_{12}$ in chloronaphthalene at 280 °C. [151] By using this method $F_{16}PcRu$ was obtained with contamination from a $F_{16}RuPc(CO)$ complex.

6.1 Synthesis of Hexadecafluoro(phthalocyaninato)ruthenium(II) 23

The synthesis and characterization of pure hexadecafluoro(phthalocyaninato)ruthenium(II), $F_{16}PcRu$ **23** is carried out by refluxing tetrafluorophthalonitrile (**22**) with $RuCl_3 \cdot 3H_2O$ in 2-ethoxyethanol for 24 hrs (Scheme 21). Complex **23** is obtained in 52% yield as an amorphous or microcrystalline black powder without any formation of the CO adduct. [151] As will be shown below **23** also has a dimeric structure.

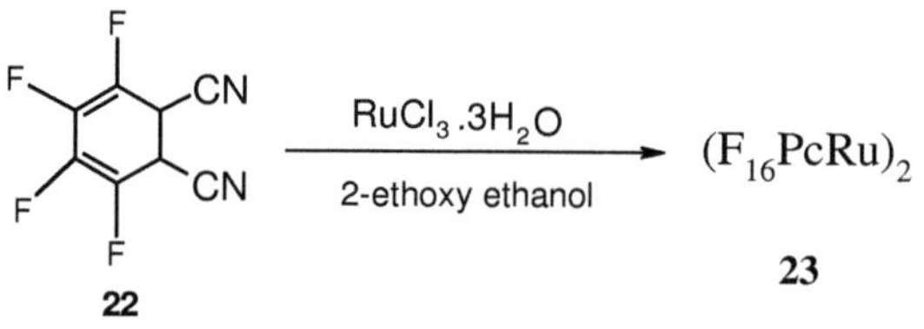

Scheme 21: Synthesis of hexadecafluoro(phthalocyaninato)ruthenium(II) (**23**)

6.2 EXAFS Spectroscopic measurements

For many phthalocyanines and napthalocyanines, it is difficult to obtain suitable single crystals for the determination of a crystal structure. In such cases, EXAFS (extended X-ray absorption fine structure) spectroscopy is considered to be a powerful technique for the determination of local structure of a specific atom, regardless of the state of the sample. EXAFS provides information on the coordination number, the nature of the scattering atoms surrounding the absorbing atom, the interatomic distance between the absorbing atom and the backscattering atoms and the Debye-Waller factor, which accounts for the disorders due to static displacements and thermal vibrations.[152, 153] (for more details see Experimental part).

To obtain more information about the structure of **23** EXAFS measurements have been carried out on amorphous **23** which are compared with the unsubstituted PcRu [141a] and $(tBu)_4PcRu$. [141b]

The experimentally determined and fitted EXAFS functions of **23** are shown in k space as well as by Fourier transforms in real space in Figure 36a, b.

The structural parameters are summarized in Table 6. The thereof deduced structure of **23** is given in Figure 37.

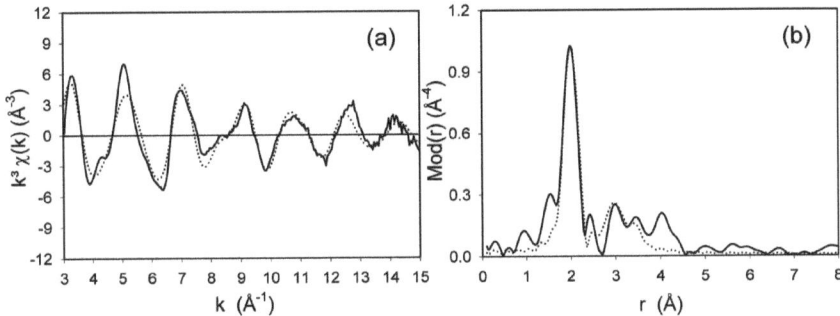

Figure 36: Experimental (solid line) and calculated (dotted line) EXAFS functions (a) and their corresponding Fourier transforms (b) for $(F_{16}PcRu)_2$ **23** measured at Ru K-edge.

Table 6. EXAFS determined structural data for $(F_{16}PcRu)_2$

A-Bs[a]	N[b]	r[c] [Å]	σ[d] [Å]	ΔE₀[e] [eV]	k-range [Å⁻¹]	Fit-Index
Ru-N$_{Indol}$	4	2.04 ± 0.02	0.059 ± 0.006	22.50	3.0 - 15.0	33.90
Ru-C	8	3.02 ± 0.03	0.122 ± 0.015			
Ru-N$_{Aza}$	4	3.32 ± 0.04	0.081 ± 0.009			
Ru-Ru	1	2.43 ± 0.03	0.122 ± 0.015			

[a] absorber (A) – backscatterers (Bs), [b] coordination number N, [c] interatomic distance r, [d] Debye-Waller factor σ with its calculated deviation and [e] shift of the threshold energy ΔE₀.

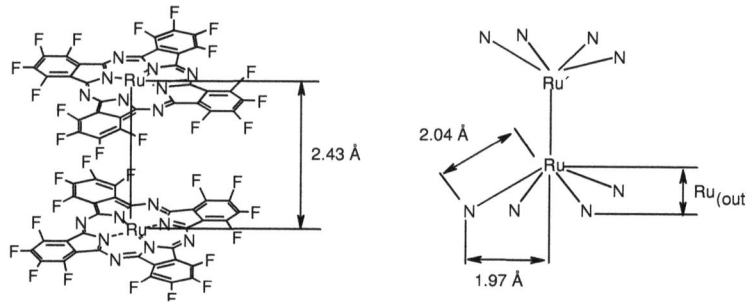

Figure 37: EXAFS-Structure of (F$_{16}$PcRu)$_2$ **23**.

In the fitting procedure, the coordination numbers were fixed to the known values of the phthalocyanine molecule and the other parameters including interatomic distances Debye-Waller factors and energy zero value were determined by iterations.

The analysis of the data shows the contribution of phthalocyanine macrocycle in the spectra. In agreement with the well-known structure of phthalocyanine complexes,[154] four nitrogen backscatterers at 2.04 Å, arising from the indol nitrogen atoms, and another four nitrogen backscatterers at 3.32 Å, stemming from the aza nitrogen atoms, were found. In addition, another shell consisting of eight carbon backscatterers at 3.02 Å was observed. All these distances can be assigned to intramolecular contribution.[154] Besides the expected three shells of the phthalocyanine macrocycle, a significant improvement in the fit index was obtained by considering a ruthenium backscatterer at a distance of 2.43 Å in the simulation of the spectrum. This distance is in good agreement with the reported ruthenium-ruthenium double bond distances in other organo-ruthenium complexes, for example 2.40 Å in (PcRu)$_2$,[146] 2.41 Å in Ru$_2$(OEP)$_2$,[155] 2.38 in Ru$_2$(C$_{22}$H$_{22}$N$_4$)[156] and 2.42 Å in tBu$_4$PcRu.[141] This result is consistent with a dimeric structure for **23**. In the Fourier transform plot a peak at about 4 Å is supposed to be due to eight carbon backscatterers in the phthalocyanine macrocycle.[147, 157] A fitting with this backscatterer in this distance range does not increase the agreement of the experimental EXAFS function compared to the fitted function, hence this shell was not considered for evaluation.

6.3 Spectroscopic characterization of 23

The UV/Vis spectrum of $(F_{16}PcRu)_2$ (**23**) recorded in dry and oxygen-free chloroform under protected conditions shows a broad Q band at $\lambda = 622$ nm, and the B-band at 434 nm.

Tabel 7 shows a comparison of the UV/Vis-maxima of **23** with PcRu [158] and $(tBu)_4PcRu$.[141] It is interesting to see that both the Q-and B-bands maxima of **23** fully substituted with electron-withdrawing fluorines and the nonsubstituted PcRu are not much different.

The electron-withdrawing effect of the fluorines in the dimeric structure of **23** therefore seems to be lower than in monomeric Pc`s.

Table 7: Q and B bands for the list of compounds.

Compound	Q(nm)		B(nm)	
$F_{16}PcRu$	704	622	434	322
PcRu [a]	710	620	435	310 [142]
$(tBu)_4$-2,3PcRu [b]	700	614	426	289 [135]

a) spectrum was recorded in fluorolube.
b) spectrum was recorded in chloroform as solvent.

6.4 Interaction of 23 with oxygen in solution

Dissolution of $(F_{16}PcRu)_2$ **23** in chloroform in the presence of oxygen or in air is accompanied by a rapid colour change of the solution from dark bluish-green to blue. Correspondingly, the UV/Vis absorption spectrum of the solution changes, as shown in Figure 38.

Initially, the spectrum shows an intense maximum Q band absorption at $\lambda = 622$ nm (spectrum 1). By recording an UV/Vis spectrum of the solution every five minutes, characteristic spectral changes are observed which go to completion within ca. one hour and determines the appearance of a new higher intensity maximum Q band absorption at $\lambda = 647$ nm (spectrum 6). This bathochromically shifts from $\lambda = 622$ to 647 nm through from spectrum 1 to spectrum 6 is assigned to interaction of $(F_{16}PcRu)_2$ **23** with atmospheric oxygen.

The phenomena described here are reproducible and were similarly described before for PcRu. [146] The structure of the reaction product of PcRu with oxygen as well as the product of $F_{16}PcRu$ with oxygen is not known.

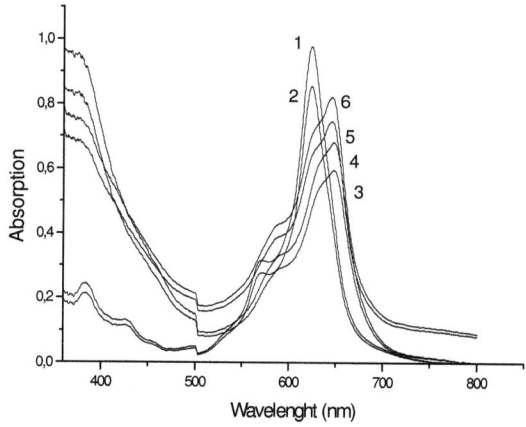

Figure 38: UV/Vis spectral variations (Curves 1-6) for solutions of $(F_{16}PcRu)_2$ **23** in CHCl$_3$ in the presence of air.

6.5 Magnetic Measurements

The temperature-dependent measurements of the magnetic susceptibility of a pure sample of $(F_{16}PcRu)_2$ **23** was carried out from 5 to ca. 300 K. The magnetic susceptibility value reach to 2.25 μ_B (T = 300K) (c.f. Figure 39), i.e. it approximates asymptotically the spin-only value of one unpaired electron (2.25 μ_B). Complex **23** therefore has only one spin per ruthenium(II) ion, although for a d^6 transition metal in a square-planar ligand field two unpaired electrons are expected. This confirms the EXAFS- results that **23** has a dimeric structure with a ruthenium-ruthenium double bond.

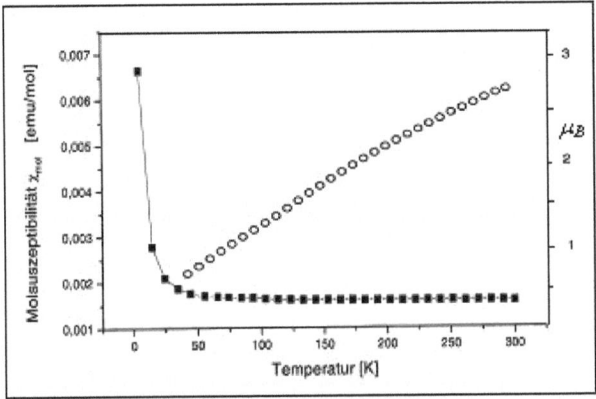

Figure 39: Magnetic moment μ (μ_B) and magnetic susceptibility χ_M (cgs) Vs T (K) for ($F_{16}PcRu)_2$ **23**.

IV. Summary

Our objective in the first section of this work is using the repetitive Diels-Alder reaction to synthesize and fully characterize the binuclear phthalocyanines **10** and **11** (Schemes 13 and 14, pages 34 and 37) [82a] and also the trinuclear phthalocyanines **13** and **20** (Schemes 16 and 20, pages 42 and 54) [82b] containing the same metal e.g. Ni **10** and **11** for the binuclear systems or Inacac for the trinuclear compound **20**.

The course of reactions for these compounds starts with the synthesis of the unsymmetrically substituted phthalocyanine **4a**, which was obtained (among the other statistical products) through the statistical condensation of phthalonitriles **1** and **2** (Scheme 11, page 29). Reaction of **4a** with tetracyclone afforded the adduct **9**, which was subsequently (via the intermediate **9a**) converted into the binuclear nickel phthalocyanine **10** (Scheme 13, page 34). Dehydration of **10** with *p*-toluenesulfonic acid (Scheme 14, page 37) gives the fully conjugated target dimer **11**.

In the second section the same approach as described for the synthesis of **10** and **11** was used to obtain the trinuclear phthalocyanine **13** (Scheme 16, page 42). The necessary starting material **5b** for preparing **13** was separated from the statistical products (Scheme 11, page 29). and reacted with tetracyclone to afford **12** (Scheme 15, page 40), which subsequently (via the intermediate **12a**) gave the planar trinuclear phthalocyanine **13** (Scheme 16, page 42)

In the third section of this work, we have demonstrated additional applications of the building block PcNi **4a** for synthesizing a series of nickel phthalocyanine **15a-c** containing furopyridine units by using pyridine *N*-oxides (Scheme 17, page 46).[98] The formation of the Pc`s **15a-c** occurs via 1,3-dipolar cycloaddition reactions as shown in Figure 23. From these results, the use of the Diels-Alder strategy as a novel and general approach to asymmetrically substituted monoadducts of nickel phthalocyanine **15a-c** has proved again to be efficient and convenient.

In the fourth section of this work (RO)$_8$PcInacac **16** (R = 2-ethylhexyl) was prepared by reacting (RO)$_8$PcH$_2$ **3b** with In(acac)$_3$ in DMF at 140°C. By the same method tetra-*tert*-butylphthalocyaninatoindium(III)acetylacetonate (*t*Bu)$_4$PcInacac **19** was obtained in good yield by reacting the metal-free (*t*Bu)$_4$PcH$_2$ **18** [113] with an excess of In(acac)$_3$ in refluxing DMF (Scheme 19). Trimer **20** was obtained from **13** by the same method (Scheme 20).

The optical limiting (OL) properties of $(RO)_8PcInacac$ **16** R = 2-ethylhexyl, $(tBu)_4PcInacac$ **19** and trinuclear Pc **20** were studied in the fifth section (see Table 5, page 63). The OL measurements were carried out in toluene at 532 nm with nanosecond laser pulses by means of the Z-scan technique (page 58).[82b] Z-scan profiles of **16**, **19** and **20** in solution have been determined. (Figure 35, page 63). Comparison of **16**, **19** and trimer **20** with the state of art molecule $(tBu)_4PcInCl$ **21a** (Figure 35, page 63) showed a larger decrease of transmittance for $(tBu)_4PcInCl$ **21a** with respect to compounds **16**, **19** and **20** in the nonlinear region.

It can be seen that the magnitude of the nonlinear absorption coefficient is largest, over the intensity region presented in the plot, for the compounds in order of **16** > **21a** > **19** > **20**. The nonlinear absorption coefficient in all cases is of the order of $10^{-8} - 10^{-7}$ cm W^{-1}.

In the last part of this work we reported an easy method for the preparation of $(F_{16}PcRu)_2$ **23** by the reaction of tetrafluorophthalonitrile with $RuCl_3.3H_2O$ in 2-ethoxyethanol (Scheme 21, page 66).[138] Dimeric $(F_{16}PcRu)_2$ **23** was obtained as an amorphous material which is soluble in common organic solvents. In spite of the 16 electronegative fluorines in the peripheral positions of the phthalocyanine ring $F_{16}PcRu$ exhibits comparable structural and magnetic properties as the earlier investigated PcRu. The dimeric structure of **23** was proved by EXAFS spectroscopy and magnetic measurements.

V. EXPERIMENTAL PART

1. General Comments

Materials

All reactions were carried out in a dry nitrogen atmosphere, unless otherwise stated. Commercially available chemicals were used as delivered by Fluka and Aldrich. All solvents were dried according to standard procedures.[159] The following precursors were prepared according to reported procedures: 4-*tert*-butylphthalonitrile [**17**], 4,5-bis(2-ethylhexyloxy)phthalonitrile [**1**], 6,7-dicyano-1,4-epoxy1,4-dihydronaphthalene [**2**] (R`= H), 6,7-dicyano1,4-epoxy-1,4-dimethyl-1,4-dihydronaphthalene [**2`**] (R`= CH_3).

1,2,3,4 Tetrafluorophthalonitrile [**22**], [Commercial source; Aldrich].

All compounds were analysed and characterized by using the following instruments:

IR-Spectroscopy

Bruker IFS 48 and Bruker Tensor 27: solid substances were grounded with KBr and pressed to pellets, liquid compounds were measured directly with Bruker Tensor 27.

UV/Vis Spectroscopy

Shimadzu UV 2102 PC. All spectra were recorded as solutions in CH_2Cl_2 or $CHCl_3$ solution. The path lengths was 1 cm.

^{1}H-NMR Spectroscopy

Bruker AC 250 (250.131 MHz): Deuterated solvents were used as an internal standard. All data are given as: chemical shift δ [ppm]. The correlation between the signals was made by using increments and by comparison with known related compounds.

^{13}C-NMR Spectroscopy

Bruker AC 250 (62.902 MHz): Deuterated solvents were used as an internal standards. All data are given as: chemical shift δ [ppm].

^{13}C-CP/MAS-NMR Spectroscopy

The ^{13}C CP/MAS (cross polarization/magic angle spinning) solid state NMR spectra were recorded on a Bruker ASX 300 multinuclear spectrometer equipped with wide bore magnets (field strength: 4.7 and 7.05 T). Magic angle spinning was applied at 10 kHz. . All data are given as : chemical shift δ [ppm] (correlated carbons).

Mass spectrometry

EI (electron impact): Finnigan TSQ 70 MAT with direct inlet, ion source temperature 200 °C, electron energy 70 eV.

FD (field desorption): Finnigan MAT 711A, temperature of the ion source: 30°C.

FAD (fast atom bombardment): Finnigan TSQ 70 MAT or Finnigan MAT 711 A. temperature of the ion source: 30°C.

MALDI-TOF (matrix assisted laser desorption ionization): Bruker, Biplex II or Bruker, Proflex II, matrix is α-cyano-cinnamic acid or 1,4-bis(phenyloxazol-2-yl)benzene (POPOP).

EXAFS Spectroscopy

The EXAFS (Extended X-ray Absorption Fine Structure) spectroscopy measurements were carried out at the research group of Prof. Dr. H. Bertagnolli, Institute of Physical Chemistry, University of Stuttgart.

EXAFS spectroscopy using a synchrotron radiation source is a useful method for probing the neighborhood environment of a selected atom in amorphous materials irrespective of the physical state of the sample. The parameters extracted include the nature of the surrounding atoms, coordination numbers, interatomic distances, and Debye-Waller factors which account for the degree of disorder (static and dynamic). [152]

The EXAFS measurements of the compound $(F_{16}PcRu)_2$ were performed at the ruthenium K-edge at 22117 eV at the beamline X1.1, at the Hamburger Synchrotron Radiation Laboratory (HASYLAB) at DESY, Hamburg, with Si(311) double crystal Monochromator under ambient conditions. The positron energy was 4.45 GeV and the beam current was about 90 mA. Data were collected in transmission mode with ion chambers filled with argon. Energy calibration was monitored with a 20-μm thick ruthenium metal foil. The sample in solid state was embedded in a polyethylene matrix and pressed into a pellet. The concentration of the solid sample was adjusted to yield an extinction of 1.5.

First, back ground absorption was removed from the experimental absorption spectrum by subtraction of a Victoreen-type polynomial. Then, the spectrum was convoluted with a series of increasingly broader Gaussian functions and the common intersection point of the convoluted spectra was taken as energy E_0. [160, 161] To determine the smooth part of the spectrum, corrected for pre-edge absorption, a piecewise polynomial was used. It was adjusted in such a manner that the low-R components of the resulting Fourier transform were minimal. After division of the background-subtracted spectrum by its smooth part, the photon energy was converted into a photoelectron wave vector scale. The resulting EXAFS function was weighted with k^3. Data analysis in k space was performed according to the curved-wave formalism of the program EXCURV92 with the XALPHA phase and amplitude functions. [162] The amplitude factor AFAC was fixed at 0.8 and an overall energy shift (ΔE_0) was introduced to give a best fit to the data. The mean free path of the scattered electrons was calculated from the imaginary part of the potential (VPI was set to –4.00).

Magnetic measurements

The temperature-dependent measurements of the magnetic susceptibility were measured with a SQUID Magnetometer (Quantum Design MPMS) between 5 and 300 K.

Elemental Analysis

C, H, N: Elementar Analysensysteme GmbH Vario EL V.

F: titration with $CeCl_3$, Murexid as indicator.

Z-Scan Measurements

The measurements of the OL properties of the synthesized compounds is also part of this work. These investigations were carried out at Molecular Electronics & Nanotechnology group, Department of Physics, Trinity College Dublin, Republic of Ireland, with the group of Prof. Dr. Werner Blau, supported by an E.U. network scientific project.

2. Synthesis

Note: the numbering of the formulas are solely for ^{1}H- and ^{13}C NMR assignments, and does not follow IUPAC rules.

2.1 Synthesis of precursors

2.1.1 1,2-Dicyano-4,5-bis(2-ethylhexyloxy)benzene (1)

2.1.1.1 1,2-Bis(2-ethylhexyloxy)benzene

Cathecol (27 g, 0.5 mol) was poured under stirring into acetonitrile (150 ml) in a three-neck flask. Then 110 ml of 2-ethylhexyl bromide (0.6 mol, 115 g) were added, followed by 85 g of K_2CO_3 (0.6 mol) and the mixture was heated and stirred at 81°C for 48 hours. The solution was allowed to cool down and filtered, to remove the base, and washed with CH_2Cl_2. The solvent was evaporated and the product chromatographed with *n*-hexane to obtain 1,2-bis(2-ethylhexyloxy)benzene as a yellow viscous oil.

Yield: 63 g (67%), yellow viscous oil.

^{1}**H NMR (CDCl$_3$):** δ = 0.89, 0.91, 0.99 (s, 12 H, CH$_3$), 1.16 (br, 2 H, CH), 1.26, 1.46, 1.60 (br, 16 H CH$_2$), 3.86, 3.93 (d, 4 H, OCH$_2$), 6.48 (dd, 2 H, H-3), 6.55 (dd, 2 H, H-4).

^{13}**C NMR (CDCl$_3$):** δ = 10.2, 13.6, (CH$_3$), 20.5, 24.7, 29.5, 31.2 (CH$_2$), 41.0(CH), 72.5(OCH$_2$), 114.5 (C-3), 125.0 (C-4), 150.0(C-1).

MS (EI, 70 eV): 335.3 [M$^+$], 222.4 [M$^+$ - C$_8$H$_{16}$].

2.1.1.2 1,2-Dibromo-4,5-bis(2-ethylhexyloxy)benzene

1,2-Bis(2-ethylhexyloxy)benzene (0.2 mol, 63 g) was poured into 250 ml CH_2Cl_2 in a three-neck flask and stirred. The solution was cooled till 0°C. Then, a solution of 0.5 mol Br_2 (76 g) in 60 ml CH_2Cl_2 was added dropwise, over 6 hours. The temperature of the reaction mixture was allowed to rise till room temperature and stirred for 2 more hours. The solution was washed with a 10% solution of NaHCO$_3$ (100 ml each time) until all the excess of the unreacted bromine was removed (3 or 4 times). Then, the organic layer was dried over MgSO$_4$ and the solvent evaporated. Column chromatography was performed with a mixture of n-hexane/CH$_2$Cl$_2$ (3/1) on silica gel to give a yellow-greenish oil.

Yield: 59.3 g, (89%), yellow-greenish oil.

^{1}H NMR (CDCl$_3$): δ = 0.98, 1.04 (m, 12 H, CH$_3$),

1.10, 1.54 (m, 16 H, CH$_2$), 1.82 (m, 2 H, CH), 3.72

(d, 4 H, OCH$_2$), 7.63 (s, 2 H, H-3).

^{13}C NMR (CDCl$_3$): δ = 10.1, 13.1 (CH$_3$), 22.6, 24.8,

28.7, 31.2 (CH$_2$), 40.3 (CH), 70.7

(OCH$_2$), 115.4 (C-4), 116.5 (C-3), 150.3 (C-1).

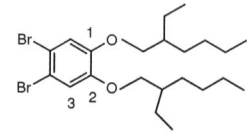

MS (EI, 70 eV): 492.6 [M$^+$], 413.4 [M$^+$ -Br], 380.0 [M$^+$- C$_8$H$_{16}$].

2.1.1.3 1,2-Dicyano-4,5-bis(2-ethylhexyloxy)benzene (1)

1,2-Dibromo-4,5-bis(2-ethylhexyloxy)benzene (0.2 mol, 100 g) was dissolved with 0.6 mol CuCN in 350 ml of freshly dried DMF. The mixture stirred and kept under reflux (155°C) for approximately 10 hours. After cooling to room temperature, a solution of 700 ml of ammonia was added to the reaction mixture, and aerated during overnight. The precipitate was filtered and washed with neutral water until no ammonia could be found in the solution, and dried in an oven at 80°C for one day. The solid was extracted with hot methanol in a Soxhlet extractor for 2-3 days. After evaporating the solvent, the dinitrile was separated by column chromatography on silica gel with a mixture of *n*-hexane/CH$_2$Cl$_2$ (3/1). A green oil, solidified after few days in the refrigerator.

Yield: **1**, 35g, (43%), green solid-oil, m.p.: 42-45°C.

^{1}H NMR (CDCl$_3$): δ = 0.85, 0.98 (m, 12 H, CH$_3$),

1.39,1.65 (m, 8 H, CH$_2$), 1.95 (m, 2 H, CH), 3.89

(d, 4 H,OCH$_2$), 7.09 (s, 2 H, H-3).

1

^{13}C NMR (CDCl$_3$): δ = 10.1, 14.8 (CH$_3$), 21.9, 22.8, 29.9, 30.8 (CH$_2$), 40.2 (CH), 70.8 (OCH$_2$), 108.8 (C-4), 116.2 (C-3), 119.2 (C-5), 153.7 (C-1).

MS (EI, 70 eV): 384.4 [M$^+$], 359.3 [M$^+$-CN], 273.0 [M$^+$-C$_8$H$_{16}$].

2.1.2 6,7-Dicyano-1,4-epoxy-1,4-dihydronaphthalene (2) and 6,7-Dicyano-1,4-epoxy-1,4-dimethyl-1,4-dihydronaphthalene (2`)

2.1.2.1 6,7-Dibromo-1,4-epoxy-1,4-dihydronaphthalene and 6,7-Dibromo-1,4-epoxy-1,4-dimethyl-1,4-dihydronaphthalene

A solution 1.6 M n-BuLi in n-hexane (25 ml, 43 mmol) was added dropwise to a solution of 15.5 g of 1,2,4,5-tetrabromobenzene (40 mmol) and 26 ml of furan or [20 ml dimethyl furan (2`)] in 350 ml of dried toluene, for 5 hours, at -15°C. The stirred mixture was allowed to come to room temperature and 4 ml of methanol was added. The solution was washed with distilled water, dried over MgSO$_4$ and the solvent was evaporated. The resulting yellowish oil was poured into a small quantity of n-hexane and recrystallized from methanol.

Yield: 6,7-Dibromo-1,4-epoxy-1,4-dihydronaphthalene, 7.0 g, (64.4%), yellow solid, m.p: 113-116°C

^{1}H NMR (CDCl$_3$): δ = 5.8 (s, 2 H, H-1), 7.1 (s, 2 H, H-2), 7.6(s, 2 H, H-6).

^{13}C NMR (CDCl$_3$): δ = 82 (C-1), 122.6 (C-7), 124.4 (C-6), 144.6 (C-5), 150.83 (C-2).

MS (EI, 70 eV): 301.8 [M$^+$]

Yield: 6,7-Dibromo-1,4-epoxy-1,4-dimethyl-1,4-dihydronaphthalene, 5.9 g, (67.2%), yellow solid, m.p.: 126°C

^{1}H NMR (CDCl$_3$): δ = 0.89 (s, 6 H, 2CH$_3$), 7.3 (s, 2 H, H-1), 7.8(s, 2 H, H-5).

^{13}C NMR (CDCl$_3$): δ = 15 (CH$_3$), 151.23 (C-1), 146.2 (C-4), 124.64 (C-5), 121.6 (C-6).

MS (EI, 70 eV): 330.8 [M$^+$]

2.1.2.2 6,7-Dicyano-1,4-epoxy-1,4-dihydronaphthalene (2) and 6,7-Dicyano-1,4-epoxy-1,4-dimethyl-1,4-dihydronaphthalene (2`)

6,7-dibromo-1,4-epoxy-1,4-dihydronaphthalene (0.03 mol, 7.9 g) or [0.03 mol of 6,7-dibromo-1,4-epoxy-1,4-dihydronaphthalene(6.8 g) (2`)] was dissolved with 0.1mol CuCN in 100 ml of freshly dried DMF. The mixture stirred and kept under reflux (155°C) for approximately 9 hours. After cooling to room temperature, a solution of 400 ml of ammonia was added to the reaction mixture, and aerated during overnight. The precipitate was filtered and washed with neutral water until no ammonia could be found in the solution, and dried in an oven at 80°C for one day. The solid was extracted with hot CH_2Cl_2 in a Soxhlet extractor for 2-3 days. After evaporating the solvent, the dinitrile was separated by column chromatography on silica gel with a mixture of n-hexane/CH_2Cl_2 (3/1). The desired product was recrystallized from $CHCl_3$ to give a white solid.

Yield: 2, 2.5 g, (39.0%) yellow solid, m.p.: 201-203°C

^{1}H NMR (CDCl$_3$): δ = 6.1 (s, 2 H, H-1), 7.63 (s, 2 H, H-2), 7.88 (s, 2 H, H-6).

^{13}C NMR (CDCl$_3$): δ = 82.8 (C-1), 112.8 (C-7), 116.2 (C-8), 124.2 (C-6), 144.5(C-5), 156.7 (C-2).

MS (EI, 70 eV): 194.1 [M$^+$], 167.9 [M$^+$-CN].

Yield: 2`, 2.0 g, (40.0%) yellow solid, m.p.: 206-208°C

^{1}H NMR (CDCl$_3$): δ = 0.89 (s, 6 H, 2CH$_3$), 7.73 (s, 2 H, H-1), 7.90 (s, 2 H, H-5).

^{13}C NMR (CDCl$_3$): δ = 15.8 (CH$_3$), 114.1 (C-6), 122.8 (C-5), 142.3(C-4), 154.8 (C-1).

MS (EI, 70 eV): 222.2 [M$^+$].

2.2 Synthesis of dienophilic phthalocyanines

2.2.1 Asymmetrically substituted mono dienophillic nickel(II)phthalocyanine 4a with AAAB-Symmetry: [2,3,9,10,16,17-Hexa(2-ethylhexyloxy)-23,26-dihydro-23,26-dimethyl-23,26-epoxybenzophthalocyaninato] nickel (4a)

A mixture of 6,7-dicyano-1,4-epoxy-1,4-dimethyl-1,4-dihydronaphthalene (**2`**) (0.9 g, 4.02 mmol), 1,2-dicyano-4,5-bis(2-ethylhexyloxy)benzene (**1**) (3.2 g, 5 mmol) and Ni(OAc)$_2$.4H$_2$O (1.10 g, 4.43 mmol) was suspended in pentanol (50 mL) in a nitrogen purged vessel and DBU (0.1 mL) was added. The mixture was heated to 145 °C and stirred for 19 h. It was allowed to cool down and poured into MeOH (200 mL). The precipitate formed was isolated using centrifugation and was washed several times with MeOH. The crude mixture of PcNi complexes was separated by flash chromatography on silica gel starting with CH$_2$Cl$_2$-hexane (4:1) as the mobile phase. After complete elution of fraction 1 to obtain **3a**, CH$_2$Cl$_2$ was used as eluent to collect **4a** as the second fraction. The solvent was removed and the blue green solids were extracted several times with acetone to achieve further purification. Drying in vacuo furnished **3a** (380 mg, 13%) and **4a** (750 mg, 27%).

IR (KBr): ν (cm^{-1}): 2958, 2927, 2871, 2858, 1606, 1463, 1425, 1386, 1303, 1205, 1138, 1107, 1070, 895, 750.

UV/ Vis (CH$_2$Cl$_2$): λ $_{max}$ = 666.50, 601.50, 403.50, 310.50, 287 nm.

^{1}H-NMR (CDCl$_3$): δ = 0.98- 1.14 (br, 36H, CH$_3$), 1.2 (br, 6 H, 3-H (2 CH$_3$)), 1.4, 1.81 (br, 48H, CH$_2$), 2.02 (br, 6H, CH), 4.38 (br, 12 H, OCH$_2$), 7.39 (s, 2H, 1H), 8.5-8.83 (2s, 6H, 10-H, 13-H, 18-H), 9.09 (s, 2 H, 5-H).

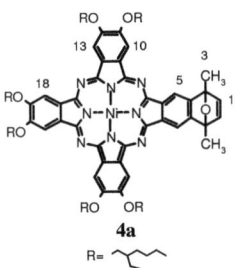

^{13}C-NMR (CDCl$_3$): δ = 11.4, 11.6, 14.3, 14.6 (CH$_3$), 21.4, 23.2, 23.4, 23.9, 24.4, 29.1, 29.5, 29.7, 30.2, 30.6, 30.9, (CH$_2$), 39.6, 39.9 (CH), 71.7 (OCH$_2$), 82.9 (C-2), 104, 104.5, 104.6 (C10, C-13, C-18), 114 (C-5), 130.3, 131(br, C-9, C-14, C-17), 135.3 (C-6), 143.4, 143.5, 144.5, 144.9 (C-1, C-4, C-8, C-16), 149.4 (C-7), 151.6, 152.3, 152.8 (C-11, C-12, C-19).

MS (FD), m/z (%): 1434.62 (100) (M$^+$).

EA: C$_{86}$H$_{118}$N$_8$NiO$_7$ (1434.62): Calcd.: C 72.00, H 8.29, N 7.81; Found: C 71.09, H 8.01, N 6.91.

2.2.2 Symmetrically substituted bis-dienophillic metal free phthalocyanine 5b with ABAB-Symmetry:[2,3,16,17-Tetra(2-ethylhexyloxy)-9,10,23,26-dihydro-9,10,23,26-epoxybenzophthalocyaninato]nickel (5b)

A mixture of 6,7-dicyano-1,4-epoxy-1,4-dihydronaphthalene (2) (3.9 mg, 20 mmol), and 1,2-dicyano-4,5-bis(2-ethylhexyloxy)benzene (1) (7.6 mg, 20 mmol) was suspended in pentanol (50 mL) in a nitrogen purged vessel and DBU (0.1 mL) was added. The mixture was heated to 145°C and stirred for 19 hrs, allowed to cool down and poured into MeOH (200 mL). The precipitate formed was isolated by centrifugation and was washed several times with MeOH. The crude mixture of the Pc complexes was separated by flash chromatography on silica gel starting with CH_2Cl_2 as the mobile phase. After complete elution of fraction 1 to obtain 3b, CH_2Cl_2-hexane (2:1) was used as eluent to collect 4b as the second fraction. After complete elution of 4b, a mixture of CH_2Cl_2-EtOAc (2:1) was used as eluent to obtain 5b as the third fraction. The solvent was removed and the blue green solids were extracted several times with acetone for further purification. Drying in vacuum furnished 3b (0.7 mg, 6 %), 4b (2.4 mg, 25 %) and 5b (2.9 mg, 26 %).

IR (KBr): ν (cm^{-1}): 3420.7 (NH), 2957, 2923, 2855, 2215, 1604, 1455, 1382, 1275, 1208, 1092, 867, 847, 744, 691, 629, 414.

UV/ Vis (CH$_2$Cl$_2$) : λ$_{max}$ = 690, 657, 641, 349 nm.

^{1}H-NMR (CDCl$_3$) : δ = -1.80 (br, 2H, NH), 1.14-1.28 (br, 24H, CH$_3$), 1.58, 1.82 (br, 32H, CH$_2$), 2.05, 2.1 (br, 4H, CH), 4.4-4.6 (br, 8H, OCH$_2$), 6.20 (br, 2H, 2-H), 7.3-7.38 (br, 4H, 1-H), 8.30, 8.4, 8.6 (br, 4H, 9-H, 12-H), 8.79, 8.87 (br, 4H, 4-H).

^{13}C-NMR (CDCl$_3$): δ [ppm] : 11.5, 11.7, 11.9, 14.4, 14.7, 14.8 (CH$_3$), 23.3, 23.6, 24.5, 29.2, 29.4, 29.7, 30.5 (CH$_2$), 39.9 (CH), 71.9 (OCH$_2$), 82.5 (C-2), 103.9 (C-9), 113.6 (C-4), 130.3 (C-8), 136 (C-5), 144 (br, C-1, C-3, C-7), 149 (C-6), 152 (C-10).

MS (FAB), m/z (%) : 1159.5 (50) M$^+$

EA: C$_{72}$H$_{86}$N$_8$O$_6$ (1159.51): Calcd. : C 74.58, H 7.47, N 9.66; Found: C 73.78, H 7.09, N 8.91

2.3 Synthesis of Ni/Ni binuclear metal-phthalocyanine 11

2.3.1 Pc-tetracyclone adduct 9

[2,3,9,10,16,17-Hexa(2-ethylhexyloxy)-23,26-dihydro-23,26-dimethyl-23,26-epoxybenzo-24,25-tetracyclone-phthalocyaninato]nickel adduct 9

A mixture of **4a** (100 mmol) and tetraphenylcyclopentadienone (39.5 mg, 102 mmol) was dissolved in toluene (30 mL) and stirred at 100 °C for 4-5 d. (TLC-control: silica gel, CHCl$_3$). The solvent was evaporated and the residue was separated by flash chromatography (CH$_2$Cl$_2$; first fraction: tetracyclone, 2nd fraction: product). Evaporation of the solvent and drying in vacuo gave an average yield of **9** (124 mg, 68%) as a blue-green solid.

IR (KBr): ν (cm^{-1}): 3078, 2959, 2928, 1711 Co, 1606, 1427, 1386, 1301, 1276, 1230, 1109, 1056, 852, 808, 750, 698.

UV/ Vis (CH$_2$Cl$_2$) : λ$_{max}$ = 673, 604, 393.50 nm.

^{1}H NMR (CDCl$_3$): δ = 0.97-1.4 (br, 42 H, CH$_3$), 1.5-1.98 (br, 48 H, CH$_2$), 2.32 (br, 6 H, CH), 3.9 (S, 2 H, 3-H), 4.46 (br, 12 H, OCH$_2$), 7.39 and 7.42 (br, 20 H, 24H-31H), 8.7 and 8.8 (3 s, br, 6H, 12-H, 15-H, 20-H), 9.33 (s,2 H, 7-H).

^{13}C-NMR (CDCl$_3$): δ = 11.4, 11.6, 14.3, 14.6 (CH$_3$), 23.2, 23.4, 24.4, 29.1, 29.7, 30.6, 30.9, (CH$_2$), 39.6, 39.9 (CH), 47.3 (C-3), 64.8 (C-2), 71.7 (OCH$_2$), 82.9 (C-4), 104, 104.5, 104.6 (C-12, C-15, C-20), 113 (C-7), 126.7, 127.4, 128.4, 128.7, 129.1, 129.7, 130.3, 131(br, C-11, C-16, C-19, C-24-31), 135.3, 135.56 (C-22), 138.6 (C-1), 146.9 (C-5, C-8, C-9, C-10, C-17, C-18), 152.3, 153.8 (C-13, C-14, C-21), 196.5 (CO).

MS (FD), m/z (%): 1819.9 (100) (M$^+$),1408.5 (40) M$^+$- Isobenzofuran

EA: C$_{115}$H$_{138}$N$_8$NiO$_8$ (1819.10): Calcd.: C 75.93, H 7.64, N 6.16; Found: C 74.75, H 6.86, N 5.78.

2.3.2 Synthesis of symmetrically substituted Ni/Ni binuclear metal-phthalocyanine 10

4a (72 mg, 0.05 mmol) and (94 mg, 0.05 mmol) of tetracyclone adduct **9** were dissolved in 30 ml of dry xylene (under nitrogen) and heated under reflux for 28 h. (TLC control, CH_2Cl_2). The solvent was evaporated and the residue was separated by flash chromatography starting with CH_2Cl_2 as eluent, whereby tetracyclone adduct **10** was eluted first in front of a green fraction. After complete elution of **4a** as the 2nd fraction a mixture of CH_2Cl_2/ethyl acetate, 4:1 was used as eluent to obtain **10** as the 3rd fraction. The solvent was removed and the blue-green solid was extracted several times with acetone to achieve further purification. Drying in vacuo furnished 36 mg (25%) of **10**.

IR(KBr): $v(cm^{-1})$: 2928, 1606, 1531, 1462, 1386, 1361, 1277, 1229, 1176, 1109, 1055, 1029, 853, 819, 751, 736 .

UV/ Vis (CH_2Cl_2): λ_{max} = 668, 396, 287 nm.

^{1}H-NMR (CDCl$_3$): δ = -0.41, -0.81, 0.88, 0.92, -0.95, 0.97, 1.08, 1.11, 1.23 (br, 72H, CH_3),1.64, 1.66 (br, 12 H, 18-H (4CH_3), -0.34 (s, 2 H,19-H), 0.81, 1.69, 1.97 (br, 96 H, CH_2), 2.00, 2.3 (br, 12H, CH), 4.39, 4.51 (br, 24 H, OCH_2), 7.4 (s, 4 H), 8.96 (s, 8 H, 2-H, 7-H), 10.01 (s, 10-H), 10.33 (br, 4 H, 15-H).

^{13}C-NMR Dept 135 (CDCl$_3$): δ = 11.4, 11.6, 11.8, 14.1, 14.4, 14.5, 16.2 (CH_3), 22.1, 22.5, 22.7, 23.1, 24.3, 30, 30.1, 30.2, 30.4, 32.1, 32.8, 33.1 (CH_2), 40.5, 40.2, 40.7 (CH),72.2 (OCH_2), 51.02 (C-19), 80.1 (C-17), 126.2 (C-2, C-7,C-10), 127.0 (C-15), 127.9, 128.6, 128.7, 128.9 (C-3, C-6, C-11, C-14, C-16), 129.04, 129.1, 130.2, 130.4 (C-4,C-5,C-12,C-13), 133.36, 133.7, 134.1 (C-1,C-8, C-9) .

MS (FD), m/z (%): 1434.62 (90) (M^+), 1410.2 (55) (M^+-Isobenzofuran), 2844.5 (20) (M^+ diepoxid dimer), 1434.62 and 1410.2 are Retro-Diels Alder fragments .

EA: $C_{170}H_{234}N_{16}Ni_2O_{14}$ (2843.21): Calcd.: C 71.81, H 8.29, N 7.88; Found: C 72.00, H 7.85, N 6.90.

2.3.3 Dehydrated binuclear metal-phthalocyanine 11

(29 mg, 0.01 mmol) of **10** was dissolved in 30 ml of freshly distilled dry toluene in a nitrogen-purged vessel and *p*-toluenesulfonic acid (50 mg, 0.26 mmol) was added. The mixture was stirred for 5 h at 80 ^{0}C. Then 1 ml of Et$_3$N was added. After 15 min. the solvent was evaporated. Subsequent flash chromatography (SiO$_2$, CH$_2$Cl$_2$) gave after drying in vacuo 6.4 mg (22%) of **11** as an olive-green powder.

UV/Vis (CH$_2$CL$_2$): λ$_{max}$ = 690.5, 673.50, 302, 286 nm.

MS (MALDI-TOF) , m/z (%) 2809.16 (100) (M$^+$).

EA: C$_{170}$H$_{232}$N$_{16}$Ni$_2$O$_{12}$ (2809.16): Calcd.: C 72.68, H 8.32, N 7.97; Found: C 71.30, H 7.80, N 6.90.

2.4 Synthesis of trinuclear metal free phthalocyanine 13

2.4.1 Pc-tetracyclone bisadduct 12

A solution of **5b** (23.2 mg, 20 mmol) and tetraphenylcyclopentadienone (38.5 mg, 100 mmol) in toluene (30 ml) was stirred at 70 °C for (3-4) days under nitrogen. After evaporation of the solvent, the crude material was purifed by preparative TLC on silica gel with chloroform in order to remove the excess of tetracyclone. The resulting green fraction was separated in two fractions by PTLC (silica gel) with a mixture of dichloromethan/ toluene (5:1). The fractions consist of the tetracyclone-bisadduct **12** and the tetracyclone-monoadduct. Yield (31.4 mg, 84 %) **12** and (2.8 mg, 9 %) tetracyclone-monoadduct. R$_f$ (silica gel, Dichloromethane): 0.8 (14), 0.51 (monoadduct) .

IR (KBr): ν (cm^{-1}): 3424 (NH), 3076, 1772 (C=O), 1652, 1558, 1496, 1417, 1398, 1350, 1271, 1218, 1159, 1105, 1074, 962, 792, 740, 678 .

UV/ Vis (CH$_2$Cl$_2$): λ$_{max}$ = 696, 661, 346 nm.

^{1}H-NMR (CDCl$_3$): δ = -0.74 (s,2H, NH), 1.11-1.36 (br, 24H, CH$_3$), 1.6-1.83 (br, 32H, CH$_2$), 2.09 (br, 4H, CH), 3.4 (br, 4H, 3-H), 4.49, 4.59 (br, 8H, OCH$_2$), 6.44 (br, 4H, 4-H), 7.1-7.52 br, 20H, 15- 18H), 7.62-7.87 (br, 20H, 19-21H), 8.44(br, 4H, 11-H), 8.88 (br, 4H, 6-H).

MS (MALDI-TOF), m/z (%) : 1928.46 (100) M$^+$

EA: $C_{130}H_{126}N_8O_8$ (1928.46): Calcd.: C 80.97, H 6.58, N 5.81; Found: C 79.30, H 7.80, N 6.90.

2.4.2 Trinuclear metal free phthalocyanine 13

A solution of the tetracyclone bisadduct **10** (38.6 mg, 20 mmol) and a three fold excess of the PcH$_2$ **5b** (69.6 mg, 60 mmol) in anhydrous toluene (25 ml) was stirred at 120°C for 2 days under nitrogen. After evaporation of the solvent, the trimer obtained was isolated and purified by Soxhlet extraction with methanol and acetone to remove TPB and excess of **5b**. Trimer **13** was obtained as a green powder (57.2 mg, 83 %).

IR (KBr): ν(cm^{-1}): 3449 (NH), 2956, 2924, 2361, 2336, 1717, 1652, 1558, 1456, 1381, 1085, 863, 702, 419.

UV/ Vis (CH$_2$Cl$_2$): λ$_{max}$ = 699, 661, 645, 343 nm.

^{1}H-NMR (THF-d$_8$): δ = -0.40, -0.46, -0.49 (br, 3H, NH), 1.16, 1.2 (br, 72H, CH$_3$), 1.71, 1.77 (br, 96H, CH$_2$); 2.25 (br, 12H, CH), 3.6 (s, 4H, 20-H), 5.33 (br, 24H, OCH$_2$), 7.1 (S, 12H, 2-H, 19-H, 21-H), 7.67 (s, 4H, 1-H), 8.41 (s, 12H, 4-H, 17-H, 23-H), 9.11 (br, 12H, 9-H, 12-H).

^{13}C-NMR (THF-d$_8$): δ =11.4, 11.8, 14.5, 14.7 (CH$_3$), 23.5, 24.1, 29.6, 31.1, 31.6 (CH$_2$), 39.1, 39.8 (CH), 51.2 (C-20), 70.4, 70.3, 72.7, 74.8 (OCH$_2$), 79.9 (C-2, C-19), 104.1, 104.6, 105.0, 105.4, 106.3 (C-9, C-12, C-17), 143.1 (C-1).

MS (MALDI-TOF), m/z (%): 3426.4 (100) M$^+$

EA: $C_{212}H_{254}N_{24}O_{18}$ (3426.46): Calcd.: C 74.31, H 7.47, N 9.8; Found: C 74.11, H 6.45, N 7.00.

2.5 Pc-pyridine *N*-oxide adducts 15a-c

A mixture of PcNi **4a** (14.34 mg, 100 mmol) and the appropriate pyridine *N*-oxide **14a-c** (50 mmol) was dissolved in the freshly distilled appropriate solvent (toluene; **14a**, xylene; **14b** and **14c**) (20 mL). The mixture was stirred at elevated temperature in an autoclave under the conditions given in Table 2 page 47. After cooling, the solvent was evaporated and the unreacted **14a-c** were separated by flash chromatography on silica gel. CH_2Cl_2 used as eluent gave fraction 1 (unreacted **14a-c**) and a mixture of CH_2Cl_2/ EtOAc (5:1) afforded fraction 2 (PcsNi **15a-c**) which was further purified by washing with methanol. Evaporation of the solvent and drying in vacuo gave the products as blue-green solids. (Table 2).

Adduct 15a

IR (KBr): ν (cm^{-1}): 2958, 2926, 2871, 2853, 1606, 1534, 1460, 1425, 1384, 1303, 1275, 1138, 1108, 1070, 895, 750, 572.

UV/ Vis (CH_2Cl_2): λ $_{max}$ = 673.50, 608.50, 413.50, 330.50 nm.

^{1}H-NMR (CDCl$_3$): δ = 0.96- 1.15 (br, 36 H, CH$_3$), 1.2 (br, 6 H, 3-H (2 CH$_3$) epoxy ring), 1.46, 1.53 (br, 48 H, CH$_2$), 2.25- 2.32 (br, 6 H, CH), 3.98 (d,1H, 7-H), 4.22 (br, 12 H, OCH$_2$), 4.61 (d, 1H, 1-H) 7.00 (d, 1H, 3-H), 8.40 (m, 1H, 5-H), 8.52- 8.74(s ,br, 4 H, 15-H, 18-H), 9.05 (s, br, 4H, 10-H, 23-H).

MS (FD), *m/z* (%): 1541.73 (100) (M$^+$) .

EA: $C_{92}H_{123}N_9NiO_8$ (1541.73): Calcd.: C 71.67, H 8.04, N 8.17; Found: C 71.64, H 8.87, N 7.60.

Adduct 15b:

IR (KBr): ν(cm^{-1}): 2959, 2925, 2871, 2858, 1606, 1481, 1427, 1386, 1303, 1201, 1138, 1104, 1020, 828, 760, 570 .

UV/ Vis (CH_2Cl_2): λ $_{max}$ = 670.50, 600.50, 403 nm.

¹H-NMR (CDCl₃): δ = 0.86- 1.15 (br, 36 H, CH₃), 1.2 (br, 6 H, 3-H (2 CH₃) epoxy ring), 1.48, 1.56 (br, 48 H, CH₂), 2.15- 2.30 (br, 6 H, CH), 4.33 (d,1H, 7-H), 4.40 (br, 12 H, OCH₂), 4.61 (d, 1H, 1-H), 7.00- 7.32 (2d, 2H, 3-H, 4-H), 8.40 (m, 1H, 5-H), 8.51-8.74 (s ,br, 4 H, 15-H, 18-H), 9.07 (s, br, 4H, 10-H, 23-H).

MS (FD), *m/z* (%)= 1527.71 (100) (M⁺) .

EA: C₉₁H₁₂₁N₉NiO₈ (1527.71): Calcd. : C 71.54, H 7.98, N 8.25; Found: C 70.99, H 6.80, N 9.01.

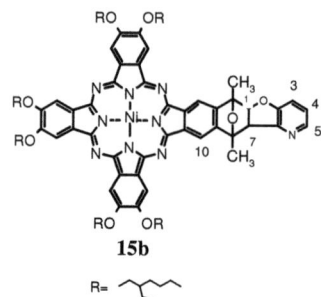

15b

R= ∧∧∧

Adduct 15c:

IR (KBr): ν(cm⁻¹): 2956, 2926, 2853, 2230 (CN),1627, 1446, 1384, 1278, 1208, 1138, 1108, 1032, 852, 747, 546 .

UV/ Vis (CH₂Cl₂): λ $_{max}$ = 669.5, 601, 405.5, 324.5 nm.

¹H-NMR (CDCl₃): δ = 0.96- 1.16 (br, 36 H, CH₃), 1.22 (br, 6 H, 3-H (2 CH₃) epoxy ring) , 1.44, 1.55 (br,48 H, CH₂), 2.15- 2.31 (br, 6 H, CH), 4.3 (d,1H, 7-H), 4.40 (br, 12 H, OCH₂), 4.62 (d, 1H, 1-H), 6.99 (d, 1H, 4-H), 8.40 (m, 1H, 5-H), 8.49-8.74 (s ,br, 4 H, 15-H, 18-H), 9.05 (s, br, 4H, 10-H, 23-H).

MS (FAB), *m/z* (%)= 1552.72 (100) (M⁺) .

EA: C₉₂H₁₂₀N₁₀NiO₈ (1552.72): Calcd.:

C 71.16, H 7.78, N 9.02 ;

Found: C 71.27, H 6.93, N 10.22.

15c

R= ∧∧∧

2.6 Synthesis of phthalocyaninatoindium(III)acetylacetonates 16, 19 and 20

2.6.1 [2,3,9,10,16,17,24,25-Octa(2-ethylhexyloxy)phthalocyaninato]indium(III) acetylacteonate (16)

A mixture of (RO)$_8$PcH$_2$ 3 (R = 2-ethylhexyl) (616 mg, 0.4 mmol) and In(acac)$_3$ (acac = acetyl acetonate) (0.06 g, 0.15 mmol) was refluxed in anhydrous DMF (15 ml) for 3 hrs under nitrogen. After cooling, methanol was added to precipitate the product. The crude product 16 was separated and purified by Soxhlet extraction with methanol several times and drying in vacuum to yield a greenish blue powder (510 mg, 73 %).

IR (KBr): ν(cm^{-1}): 2925, 1726 (C=O), 1609, 1470, 1429, 1389, 1217, 1180, 752.

UV/ Vis (CH$_2$Cl$_2$): λ$_{max}$ = 696, 627, 404, 350 nm.

^{1}H NMR(CDCl$_3$): δ = 0.80 (br, 6H, methyl groups of acac), 1.2-1.3 (br, 48H, CH$_3$), 1.54 (br, 64H, CH$_2$), 1.79 (br, 9H, CH), 4.15 (br, 16H, OCH$_2$), 8.92 (s, 8H, 2-H, 7-H, 10-H, 15-H).

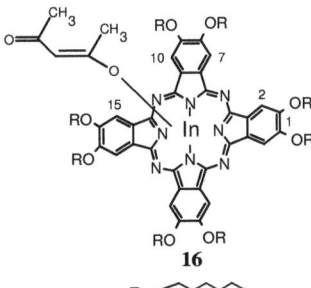

MS (FD), m/z (%) : 1748.12 (100) M$^+$

EA: C$_{101}$H$_{153}$N$_8$O$_{10}$In (2809.16) :Calcd. C 69.15, H 8.79, N 6.38; Found: C 71.10, H 8.87, N 7.20.

2.6.2 [2,(3)-Tetra-*tert.*-butylphthalocyaninato]indium(III)acetylacetonate(19)

A mixture of (*t*Bu)$_4$PcH$_2$ (18) (0.73 g, 1 mmol) and In(acac)$_3$ (acac = acetyl-acetonate) (0.12 g, 0.3 mmol) was refluxed in anhydrous DMF (15 ml) for 2 hrs under nitrogen. After cooling, methanol was added to precipitate the product. The crude Pc 19 was separated and purified by Soxhlet extraction with methanol several times and drying in vacuo to yield a green powder (0.8 g, 84 %).

IR (KBr): ν(cm^{-1}): 2959, 1726 (C=O), 1589, 1488, 1391, 1279, 1086, 831, 761, 672, 500.

UV/ Vis (CH$_2$Cl$_2$): λ$_{max}$ = 671, 607, 410, 361 nm.

^{1}H-NMR (THF-d$_8$): δ = 0.90 (br, 6H, methyl groups

of acac), 1.4-1.72 (m, 36H, *t*Bu), 2.8 (s, 1H, CH), 8.34-8.38 (m, 4H, 1-H), 9.41-9.44 (m, 4H, 2-H), 9.46-9.6 (m, 4H, 2´-H).

MS (FD), m/z (%) : 950.8 (100) M$^+$

EA: $C_{53}H_{55}N_8O_2In$ (950.88): Calcd.: C 66.94, H 5.83, N 11.78; Found: C 71.01, H 6.05, N 12.08.

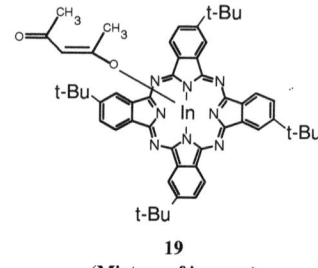

19
(Mixture of isomers)

2.6.3 Symmetric trinuclear phthalocyaninatoindium(III)acetylacetonate 20

A mixture of trimer **15** (17.1 mg, 5 mmol) and In(acac)$_3$ (acac = acetylacetonate) (0.18 g, 0.4 mmol) was refluxed in anhydrous DMF (15 ml) for 3 hrs under nitrogen. After cooling, methanol was added to precipitate the product and purificated by Soxhlet extraction with methanol and ethyl acetate to remove the rest of the metal-free trimer **15**. Trimer **20** was obtained as a blue green powder (17.3 mg, 84 %).

IR (KBr): ν(cm^{-1}): 2925, 1726 (C=O), 1455, 1383, 1377, 1282, 1217, 1119, 752.

UV/ Vis (CH$_2$Cl$_2$): λ$_{max}$ = 686, 420, 350 nm .

^{13}C CP/MAS NMR: δ = 10.68 (CH$_3$), 24.42(CH$_2$), 29.47(CH$_2$), 40(CH), 51(C-20), 73.71(C-2, C-OCH$_2$), 83.64(C-2, C-19), 108.61(C-31, C-28, C-12, C-9, C-4), 124.18 (C-23, C-17), 149.72 (C-32, C-27, C-24, C-22, C-18, C-16, C-11, C-8, C-5, C-3,C-1), 166.06 (C-11, C-10, C-29, C-30), 79.2 (C-33, C-26, C-25, C-15, C-14, C-7, C-6),193.4 (C=O).

MS (MALDI-TOF), m/z (%) : 4062.24 (100) M$^+$

EA: $C_{227}H_{269}N_{24}O_{24}In_3$ (4062.24):Calcd.: C 67.12, H 6.67, N 8.27; Found: C 68.30, H 6.15, N 9.90 .

2.7 Synthesis of Hexadecafluoro(phthalocyaninato)ruthenium(II); $(F_{16}PcRu)_2$ (23):

To a solution of tetrafluorophthalonitrile (Aldrich) (**22**; 1.7g, 8.5 mmol) in refluxing 2-ethoxyethanol (10ml) was added a solution of $RuCl_3 \cdot 3H_2O$ (0.5 g, 0.17 mmol) in the same solvent (10ml). The mixture was refluxed for 24 hours under a stream of N_2. A dark green solution was formed, which was cooled to room temperature and then poured into $MeOH/H_2O$ (3:1). The precipitate was centrifuged and dried. The crude dimeric $(F_{16}PcRu)_2$ **23** obtained was heated at 80 °C under a vacuum to sublime any unreacted tetrafluorphthalonitrile (**22**). Further purification was carried out by column chromatography over silica gel using a mixture of n-hexane/ acetone 2.5:1. After evaporation of the solvent and drying at (50°C, 0.01 Torr) a dark bluish-green powder was obtained. Yield: 0.95 g (52%).

IR (KBr): $\nu(cm^{-1})$: = 3116 cm^{-1}, 3050, 2914, 1623, 1554, 1490, 1451, 1320, 1270, 1170, 960, 840, 810, 620.

UV/ Vis (Acetone): λ_{max} = 622, 583, 434, 322 nm.

MS(FAB), m/z (%): 1802.90 (M^+) , 901.4 (M^+- monomeric form).

EA: $C_{32}N_8F_{16}Ru$ (901.4): Calcd.: C, 42.64; N, 12.43; F, 33.72 %. Found: C, 43.1; N, 11.22; F, 32.55%.

VI. Literature

[1] a) C.C. Leznoff, A.B.P. Lever (Eds.), in *Phthalocyanine:, Properties and Applications*, Vol. 1-4, VCH, New York, **1989-1996**.

[2] a) M. Casstevens, M. Samok, J. Pfleger, P. N. Prasad, *J. Chem. Phys.* **1990**, 92, 2019.

b) J. Simon, P. Bassoul, S. Norvez, *New J. Chem.* **1989**, 13, 13.

[3] a) J.F. Van der Pol, E. Neeleman, J. W. Zwikker, R. J. M. Nolte, W. Drenth, J. Aerts, R Visser, S.J. Picken, *Liq. Cryst.* **1989**, 6, 577.

b) J. Simon C. Sirlin, *Pure Appl. Chem.* **1989**, 61, 1625.

[4] a) G.G. Roberts, M.C. Petty, S. Baker, M.T. Fowler, N.J. Thomas, *Thin Solid Films* **1985**, 132, 113.

b) M.J. Cook, A. J. Dunn, F.M. Daniel, R.C.O. Hart, R.M. Richardson, S.J. Roser, **1988**, 159, 395.

c) S. Palacin, P. Lesieur, I. Stefanelli, A. Barraud, *ibid* **1988**, 159, 83.

d) M.A. Mohammad, P. Ottenbreit, W. Prass, G. Schnurpfeil, D. Wöhrle, *ibid* **1992**, 213, 285.

[5] H. Schultz, H. Lehman, M. Rein, M. Hanack, *Struct. Bonding (Berlin)* **1991**, 74, 41.

[6] J.E. Kuder, *J. Imag. Sci.* **1988**, 32, 51.

[7] a) M.-T. Riou, C. Clarisse, *J. Electroanal. Chem.* **1988**, 249, 181.

b) D. Schlettwein, D. Wöhrle, N. I. Jaeger, *J. Electrochem. Soc.* **1989**, 136, 2882.

[8] a) M. Hanack, A. Datz, R. Fay, K. Fischer, U. Keppeler, J. Koch, J. Metz, M. Mezger, O. Schneider, H.-J. Schulze, in *Handbook of Conducting Polymers* (Ed. T. A. Skotheim), Vol.1, Mercel Decker, New York **1986**.

b) M. Hanack, S. Deger, A. Lange, *Coord. Chem. Rev.* **1988**, 83, 115.

[9] a) T.J. Marks, *Science.* **1985**, 227, 881.

b) T.J. Marks, *Angew. Chem. Int. Ed. Engl.* **1990**, 29, 857.

c) B. M. Hoffman, J.A. Ibers, *Acc. Chem. Res.* **1983**, 16, 15.

[10] K. Abe, H. Saito, T. Kimura, Y. Ohkatsu, T. Kusano, *Macromol. Chem.* **1989**, 190, 2693.

[11] a) R.A. Collins, K.A. Mohamed, *J. Phys. D* **1988**, 21, 154.

b) T.A. Temofonte, K.F. Schoch, *J. Appl. Phys.* **1989**, 65, 1350.

c) Y. Sadaoka, T.A. Jones, W. Göpel, *Sensors Actuators B* **1990**, 1, 148.

[12] M. Kato, Y. Nishioka, K. Kaifu, K. Kawamura, S. Ohno, *Appl. Phys. Lett.* **1985**, 86, 196.

[13] K.-Y. Law, *Chem. Rev.* **1993**, 93, 449.

[14] J. R. Darwent, P. Douglas, A. Harriman, G. Porter, M. C. Richoux, *Coord. Chem. Rev.* **1982**, 44, 83.

[15] D.Wöhrle, J. Gitzel, G. Krawczyk, E. Tsuchida, H. Ohno, T. Nishisaka, *J. Macromol. Sci. Chem. A* **1988**, 25, 1227.

[16] B. A. Henderson, T. J. Dougherty, *Photochem. Photobiol.* **1992**, 55, 145.

[17] a) D. Wöhrle, A. Ardeschirpur, A. Heuermann, S. Müller, G. Graschew, H. Rinneberg, M. Kohl, J. Neukammer, *Makromol. Chem., Makromol. Symp.* **1992**, 59, 17.

b) D. Wöhrle, M. Shopova, S. Müller, A. D. Milev, V. N. Mantareva, K. K. Krastev, *Photochem. Photobiol. B.* **1993**, 21, 155.

[18] T.J. Klofta, J. Danzinger, P. Lee, J. Pankow, K.W. Nebesny, N.R. Armstrong, *J. Phys. Chem.* **1987**, 91, 5646.

[19] D. Schlettwein, M. Kaneko, A. Yamada, D. Wöhrle, N.I. Jaeger, *J. Phys. Chem.* **1991**, 95, 1748.

[20] J.J. Simon, H.J. Andre, *Molecular Semiconductors*, Sprinter, Berlin, **1985**, 243.

[21] D. Wöhrle, D. meissner, *Adv. Mater.* **1991**, 3, 129.

[22] R.O. Loutfy, C.K. Hsiao, A.M. Hor, G.J. Dipaola-Baranyl, *J. Imaging Sci.* **1985**, 29, 148.

[23] S. Takano, T. Enokida, A. Kabata, *Chem. Lett.* **1984**, 2037.

[24] a) R.P. Linstead, *Br. Ass. Adv. Sci. Rep.*, **1933**, 465.

b) R.P. Linstead, *J. Chem. Soc.* **1934**, 1016.

c) H. de Diesbach, E. von der Weid, *Helv. Chim. Acta.* **1927**, 10, 886

d) J.M. Robertson, *J. Chem. Soc.* **1935**, 615.

e) J.M. Robertson, *J. Chem. Soc.* **1936**, 1195.

[25] a) K. Kasuga, M. Tsutsui, *Coord. Chem. Rev.* **1980**, 32, 67.

b) A.B.P. Lever, M. R. Hempstead, C.C. Leznoff, W. Liu, M. Melnik, W.A. Nevin, P. Seymour, *Pure Appl. Chem.* **1986**, 58, 1467.

[26] a) A. Gieren, W. Hoppe, J. Chem. Soc. *Chem. Commun.* **1971**, 413.

b) W.E. Bennett, D.E. Broberg, N.C. Baenziger, *Inorg. Chem.* **1973**, 12, 930.

c) T. Kobayashi, *Bull. Inst. Chem. Res., Kyoto* **1978**, 56, 204.

d) H. Sugimoto, T. Higashi, M. Mori, *Chem. Lett.* **1983**, 1167.

[27] Hu.Te. Tseng, Hu. Tsai-Wei; Liu. Heng-Ta, Li. Jung-Chang; Li. Chung-Chun; Chen. Ming-Chia, *Japanese Patent* 20010713, **2003**.

[28] P. Gregory, *J. Porphyrins Phthalocyanines.* **2000**, 4, 432.

[29] C.H. Yang, S.F. Lin, H.L. Chen, C. Chang, *Inorg.Chem.* **1980**, 19, 3541.

[30] M. Hanack, M. Lang, *Adv. Mater.* **1994**, 6, 11, 819.

[31] M. Hanack, G. Schmid, M. Sommerauer, *Angew. Chem.* **1993**, 105, 1540; *Angew. Chem. Int. Ed. Engl.* **1993**, 32, 1422.

[32] M. Hanack, M. Sommerauer, *J .Am. Chem. Soc.* **1996**, 118, 10085.

[33] M.J. Cook, M.F. Daniel, K.J. Harrison, N.B. McKeown, A.J. Thomson, *J. Chem. Soc. Chem. Commun.* **1987**, 1148.

[34] G. Schmid, M. Sommerauer, M. Geyer, M. Hanack, C.C. Leznoff, A.B.P. Lever (Eds.),in *Phthalocyanine:, Properties and Applications*, Vol. 4, VCH, New York, **1996**,1.

[35] B. Hauschel, R. Jung, M. Hanack, *Eur. J. Inorg. Chem.* **1999**, 693.

[36] R. W. Murray, *Acc. Chem. Res.* **1980**, 13, 135.

[37] J.G. Young, W. Onyebuagu, *J. Org. Chem.* **1988**, 55, 2155.

[38] G. Manecke, D. Wöhrle, *Makromol. Chem.* **1968**, 120, 192.

[39] R. Müller, D. Wöhrle, *Makromol. Chem.* **1975**; 176, 2775.

[40] a) A.J. Epstein, B.S. Wildi, *J. Chem. Phys.* **1960**, 32, 324.

b) D. Wöhrle, *Adv. Polm. Sci.* **1983**, 50, 46.

[41] M. Rack, M. Hanack, *Angew. Chem.* **1994**, 106, 1712.

[42] M. Hanack, P. Stihler, *Eur. J. Org. Chem.* **2000**, 303.

[43] K. Ukei. *Acta Cryst.* **1973**, B 29, 2290.

[44] M. Kimura, K. Wada, K. Ohta, K. Hanabusa, H. Shirai, N. Kobayashi. N. *Macromol.* **2001**, 34, 4706.

[45] T. Zipplies, M. Hanack. *Methoden der Organische Chemie.* G. Thieme-Verlag: Stuttgart, **1987**, Vol. E 20, 2237.

[46] E. Ciliberto, K.A. Doris, W.J. Pietro, G.M. Resiner, D.E. Ellis, I. Frigala, F. H. Herbstein, M. A. Rantner, T. J. Marks, *J. Am. Chem. Soc.* **1984**, 106, 7748.

[47] D.W. DeWulf, J.K. Leland, B.L. Wheeler, A.J. Bard, D.A. Batzel, D.R. Dininny, M. Kenney, *Inorg. Chem.* **1987**, 26, 266.

[48] J. B. Khurgin, *In Nonlinear Optics in Semiconductors II*: Academic Press: San Diego, **1990**, Vol: 59, 1.

[49] P.A. Franken, A.E. Hills, C.W. Peters, and G. Weinreich, Phys. *Rev. Lett.*, **1961**, 7,118.

[50] P. A. Miles, *Appl. Opt.* **1994**, 33, 6965.

[51] T. Xia, D. J. Hagan, A. Dogariu, A. A. Said, and E. W. Van Stryland, *Appl. Opt.* **1997**, 36, 4110.

[52] D.J. Hagan, T. Xia, A.A. said, T.H. Wei, E.W. Stryland, Int. *J. Nonlin. Opt. Phys.* **1993**, 2, 483.

[53] J.S. Shirk, R.G.S. Pong, S. R. Flom, F.J. Bartoli, J.R. Lindle, M.E. Boyle, R.F. Cozzens, J. D. Adkins, A. W. Snow, *Proceedings of the Second DOD Workshop on Optical Limiters*, **1994**, 609.

[54] M. Sheik-Bahae, A.A. Said, T. Wei, D.J. Hagan, E.W. Van Stryland, *IEEE J. Quantum Electron.* **1990**, 26, 760.

[55] M. Sheik-Bahae, A.A. Said, D.J. Hagan and E.W. Van Stryland, *Opt. Lett.* **1989**, 14, 955.

[56] M. Hercher, *Appl. Opt.* **1967**, 6, 94.

[57] B.L. Justus, A.J. Campillo, A.L. Huston, *Opt. Lett.* **1994**, 19, 673.

[58] J. Robertson, P. Milsom, J. Duignan, G. Bourhill, *Opt. Lett.* **2000**, 25, 1258.

[59] K. Mansour, M.J. Soileau, E. W. Van Stryland, *J. Opt. Soc. Am. B*, **1992**, 9, 1100.

[60] K.M. Nashold, D.P. Walter, *J. Opt. Soc. Am. B*, **1995**, 12, 122.

[61] H. S. Nalwa, J. S. Shirk, *In Phthalocyanines: Properities and Applications*. C. C. Leznoff, C. C., A. J. P. Lever, Eds., VCH: Weinhein, **1996**, Vol.4, 79.

[62] M.A. Kramer, W.R. Tompkin, R.W. Boyed. *Phys. Rev.* **1986**, A 34, 2026.

[63] N. Kobayashi, *Coord. Chem. Rev.* **2002**, 227, 129.

[64] H. Engelkamp, S. Middlebeek, R.J.M. Nolte, *Sceince*, 284, **1999**, 785.

[65] N. Kobayashi, W.A. Nevin, *Chem. Lett.* **1998**, 851.

[66] N. Kobayashi, R. Higashi, B.C. Titeca, F. La mote, A. Ceulemans, *J. Am. Chem. Soc.* 121, **1999**, 1208.

[67] N. Kobayashi, A. Muranaka, *Chem. Commun.* **2000**, 1855.

[68] a) J. W. Perry, K. Mansour, I.-Y. S. Lee, X.-L. Wu, P. V. Bedworth, C.-T. Chen, D. Ng, S.R. Marder, P. Miles, T. Wada, M. Tian, H. Sassabe, *Science*, **1996**, 273, 1533.

[69] X. Wang, C.-L. Liu, Q.-H. Gong, Y.-Y. Huanh, C.-H. Huang, J.-Z. Jiang, *Appl. Phys. A* **2002**, 497.

[70] Y. Chen, L.R. Subramanian, M. Barthel, M. Hanack, *Eur. J. Inorg. Chem.* **2002**, 1032.

[71] Y. Chen, L.R. Subramanian, M. Fujitsuka, O. Ito, S. O'Flaherty, W.J Blau,T.Schneider, D. Dini, M. Hanack, *Chem. Eur. J.* **2002,** 8, 4248.

[72] H. Huckstadt, H. Homborg, *Z. Anorg. Allg. Chem.* **1997**, 369.

[73] H. Huckstadt, C.Bruhn, H. Homborg, *J. Porphyrins Phthalocyanines.* **1997**, 367.

[74] M. Gorsch, H. Homborg, *Z. Anorg. Allg. Chem.* **1998**, 634.

[75] N. Kobayashi, F. Furuya, G. C. Yug, *J. Porphyrins Phthalocyanines.* **1999**, 433.

[76] N. Kobayashi, F. Furuya, G. C. Yug, H. Wakita, M. Yokomizo, N. Ishikawa,

Chem. Eur. J. **2002**, 8.

[77] Y. Chen, M. Barthel, M. Seiler, L.R. Subramanian, S. Vagin, M. Hanack, *Angew. Chem.* **2002**, 114, 3373; *Angew. Chem. Int. Ed.* **2002**, 41, 3239-42.

[78] Y. Chen, S. O'Flaherty, M. Fujitsuka, L.R. Subramanian, O. Ito, W.J Blau, M. Hanack, *Adv. Mat.* **2003**, 15, 899.

[79] M. Barthel, D. Dini, S. Vagin, M. Hanack, *Eur. J. Org. Chem.* **2002**, 3756.

[80] R. Jung, PhD Thesis, Univ. Tübingen, Germany **2000**

[81] A-D. Schlüter, *Adv. Mater.* **1991**; 3: 282.

[82] a) T. E. Youssef, M. Hanack, *J. Porphyrines & Phthalocyanines.* **2002**, 6, 571.

b) T. E. Youssef, M. Hanack, *Eur. J. Org. Chem* **2004**, 1, 101.

[83] M Rack, B. Hauschel, M. Hanack, *Chem. Ber.* **1996**, 129, 237.

[84] R. Jung, K.-H. Schweikart, M. Hanack, *Eur. J. Org. Chem.* **1999**, 11, 646.

[85] T. Linssen, M. Hanack, *Chem. Ber.* **1994**; 127, 2051.

[86] N. McKeown, I. Chambrier, M. Cook, *J. Chem. Soc. Perkin Trans.*, **1990**, 1, 1169.

[87] D. Ruff, S. Fiedler, M. Hanack, *Synth. Met.* **1995**, 69, 579.

[88] M. Hanack, A. Gül, A. Hirsch, B. K. Mandal, L. R. Subramanian, E. Witke, *Mol.Cryst. Liq. Crysr.* **1990**, 187, 365.

[89] A.B.P. Lever, *Adv. Inorg. Radiochem.* **1965**, 7, 27.

[90] A. M. Schaffer, M. Gouterman, *Theor. Chim. Acta* **1972**, 25, 62.

[91] A. M. Schaffer, M. Gouterman, E. R. Davidson *Theor. Chim. Acta* **1973**, 30, 9.

[92] G. Torre, M. V. Martinez-Diaz, P. R. Ashton, T. Torres, *J. Org. Chem.* **1998,** 63, 8888.

[93] G. Torre, M. V. Martinez-Diaz, T. Torres, *J. Porphyrins Phthalocyanines*, **1999**, 3, 560.

[94] M. J. Cook, M. J. Heeney, *Chem Commun.* **2000**, 969.

[95] R. Jung, K.-H. Schweikart, M. Hanack, *Synth. Met.* **2000**, 111, 453

[96] B. Behnisch, P. Martinez-Ruiz, K.-H. Schweikart, M. Hanack, *Eur. J. Org. Chem.* **2000**, *14*, 2541.

[97] P. Stihler, B. Hauschel, M. Hanack, *Chem. Ber.* **1997**, 130, 801.

[98] T.E. Youssef, M. Hanack, *J. Porphyrines & Phthalocyanines.* **2005**, 9, 28.

[99] R. Jung, M. Hanack, *Synthesis.* **2001**,1386.

[100] a) B. Hauschel, B. Ruff, M. Hanack. *J. Chem. Soc. Chem. Commun.* **1995**, 2449.

b) B. Hauschel, P. Stihler, M. Hanack. *Polym. Sci.* **1996**, 4, 348.

[101] M. Podesta, D. Aubert, J.C. Ferrand. *Eur. J. Med. Chem.* **1974**, 9, 487.

[102] M. Yasuda, M.K Harano, K.Kanematsu. *J. Org. Chem.* **1981**, 46, 3836.

[103] K. Harano, R. Kondo; M. Murase; T. Matsuoka, T. Hisano, *Chem. Pharm. Bull.* **1986**; 34, 966.

[104] T. Hisano, K. Harano, T. Matsuoka, H. Yamada, M. Kurihara. *Chem. Pharm. Bull.* **1987**, 35, 1049.

[105] T. Hisano, K. Harano, T. Matsuoka, T. Suzuki, Y. Murayama. *Chem. Pharm. Bull.* **1990**, 38, 605.

[106] H. S. Nalwa, A. Kakuta *Thin Sol. Films* **1995**, 254, 218.

[107] D. Dini, M. Barthel, M. Hanack, *Eur. J. Org. Chem.* **2001**, 3759.

[108] D. Dini, M. Hanack, Physical Properties of Phthalocyanine-based Materials, *The Porphyrin Handbook,* vol 17, 107, 1 (Eds.: K. A. Kadish, K. M. Smith, R. Guilard) Academic Press, New York, **2003**.

[109] K.P. Unnikrishnan, J. Thomas, V.P.N. Nampoori, C.P.G. Vallabhan, *Opt.Comm.* **2003**, 217, 269.

[110] J.P. Linsky, T. R. Paul, R.S. Nohr, Me. Kenney, *Inorg. Chem.* **1980**, 19, 3131.

[111] T.C.Wen, I.D. Lian, I. D. *Synth. Met.* **1996**, *83*, 111.

[112] K. Varmuza, G. Maresch, A. Meller, *Mh.Chem.***1974**, 105, 327.

[113] S.A. Mikhalenko, S.V. Barkanova, O.L. Lebedev, E.A. Lukyanets, *Zh. Obshch.Khim.* **1971**, 41, 2735; *J. Gen. Chem. USSR.* **1971**, 41, 2770.

[114] M. Hanack, J. Metz, G. Pawlowski, *Chem Ber.* **1982**, 115, 2836.

[115] T. Schneider, H. Heckmann, M. Barthel, M. Hanack, *Eur. J. Org. Chem.* **2001**, 16, 3055.

[116] L.W. Tutt, T.F. Boggess, *Progr. Quant. Electr.* **1993**, 17, 299.

[117] K. Abe, H. Saito, T. Kimura, Y. Ohkatsu, T., *Macromol. Chem.* **1989**, 190, 2693.

[118] R.A. Collins, K.A. Mohamed, *J. Phys. D.*, **1988**, 21, 154.

[119] M. Kato, Y. Nishioka, K. Kaifu, K. Kawamura, S. Ohno, *Appl. Phys. Lett.* **1985**, 86, 196.

[120] M. Hanack, T. Schneider, M. Barthel, J.S. Shirk, S.R. Flom, R. G. S. Pong, *Coord. Chem. Rev.* **2001**, 219-221, 235.

[121] a) J. S. Shirk, R.G.S. Pong, S.R. Flom, H. Heckmann, M. Hanack, *J. Phys. Chem. A* **2000**, 104.

b) H.S. Nalwa, S. Miyata, In Nonlinear Optics of Organic Molecules and Polymers, *CRC Press*: Boca Raton, FL, **1997**, 813.

[122] J.S. Shirk, R.G. Pong, F.J. Bartoli, A.W. Snow, *Appl.Phys. Lett.* **1993**, 63, 1880.

[123] M. Hanack *Abstracts of Papers, 222nd ACS National Meeting*, Chicago, IL, United States, August 26-30, **2001**, POLY-006.

[124] M. Hanack, H. Heckmann, *Eur. J. Inorg. Chem.* **1998**, 3, 367.

[125] G. Rojo, G. Martin, F. Agullo-Lopez, T.Torres, H. Heckmann, M. Hanack, *J. Phys. Chem. B* **2000**, 104, 7066.

[126] C.R. Mendonca, L.Gaffo, L. Misoguti, W.C. Moreira, O.N. Oliveira, S.C. Zilio, *Chem. Phys. Lett.* **2000**, *323*, 300.

[127] F. Kajzar, J. Messier, C. Rosilio, *J. Appl.Phys.* **1986**, 60, 3040.

[128] G. Assanto, *J. Mod. Opt.* **1990**, 37, 855.

[129] *Optical Phase Conjugation* (Ed.: R. A. Fischer), Academic Press: New York, **1984**

[130] R.C. Lind, D.G. Steel, G.J. Dunning, *Opt. Eng.*, **1982**, 21, 190.

[131] A. Yariv, *IEEE J. Quant. Electr.*, **1978**, QE-14, 650.

[132] B.F. Levine, C.G. Bethea, *Appl. Phys. Lett.*, **1974**, 24, 445.

[133] E. W. van Stryland, M. Sheik-Bahae, A. A. Said, D.J. Hagan, *Prog. Cryst. Growth Charact.*, **1993**, 27, 279.

[134] J.F. Reintjes, in *Nonlinear Optic Parametric Processes in Liquids and Gases*, Academic Press: New York, **1984**, 327.

[135] J. S. Shirk, J. R. Linde, F. J. Bartoli, Z. H. Kafafi, A. W. Snow in *Materials for Nonlinear Optics – Chemical Perspectives* (Eds : S. R. Marder, J. E. Sohn, G. D. Stucky), ACS Symposium series 455 **1991**, 626.

[136] H. S. Nalwa, M. Hanack, G. Pawlowski, M. K. Engel *Chem. Phys.* **1999**, 245, 17.

[137] Y. Chen, M. Fujitsuka, S. O'Flaherty, M. Hanack, O. Ito, and W. J. Blau, *Adv. Mater.* **2003**, 15, 899.

[138] T. E. Youssef, V. Krishnan, H. Bertagnolli and M. Hanack., *unpublished results*.

[139] W. Kobel, M. Hanack, *Inorg. Chem.* **1986**, 25, 103.

[140] R. Polley, M. Hanack, *Synthesis* **1997**, 3, 295.

[141] a) M. Hanack, S. Knecht, R. Polley, *Chem. Ber.* **1995**, 128, 929.

b) A.Weber, T.S. Ertel, U. Reinöhl, H. Bertagnolli, M. Leuze, M. Hees, M. Hanack, *Eur. J. Inorg. Chem.* **2000**, 2289.

[142] M. Hanack, R. Polley, *Inorg. Chem.* **1994**, 33, 3201.

[143] B.D. Rihfer, M.E. Kenny, W.E. Ford, M.A. Rodgers, *J. Am. Chem. Soc.* **1990**, 112, 8064.

[144] M. Hanack, J. Osio-Barcina, E. Witke, J. Pohmer, *Synthesis* **1992**, 211.

[145] M. Hanack, M. Hees, E. Witke, *New J. Chem.* **1998**, 169.

[146] A. Capobianchi, A. M. Paletti, G. Pennesi, G. Rossi, R. Caminiti, C. Ercolani, *Inorg. Chem.* **1994**, 33, 4635.

[147] H. Bertagnolli, A. Weber, W. Hörner, T.S. Ertel, U. Reinöhl, M. Hanack, M. Hees, R. Polley, *Inorg. Chem.* **1997**, *36*, 6397.

[148] F.A. Cotton, N.F. Curtis, C.B. Harris, B.F. G Johnson, S.J. Lippard, J. T. Mague, W. R. Robinson, J.S. Wood, *Science.* **1964**, 145, 1305.

[149] F.A. Cotton, *Inorg. Chem.* **1965**, 4, 334.

[150] T.E. Youssef, *unpublished results.*

[151] a) K.J. Balkus, M. Eissa, R. Lavado, *Studies in Surfance Science and Catalysis*, **1995**, 94, 713.

b) K.J. Balkus, M. Eissa, R. Lavado, *J. Am. Chem.* Soc. **1995**, 117, 10753.

[152] H. Bertagnolli, T.S. Ertel, *Angew. Chem. Int. Ed.* **1994**, 33, 45.

[153] F.W. Lytle, D. E. Sayers, E. A. Stern, *Phys. Rev. B.* **1975**, 11, 4825.

[154] Jr.L.H. Vogt, A. Zalkin, D. H. Templeton, *Inorg. Chem.* **1967**, 6, 1725.

[155] J.P. Collman, C. E. Barnes, P. N. Swepston, J.A. Ibers, *J. Am. Chem. Soc.* **1984**, 106, 3500.

[156] L.F. Warren, V. L. Goedken, *J. Chem. Soc.* **1978**, 909.

[157] A. Weber, T.S. Ertel, U. Reinöhl, M. Feth, H. Bertagnolli, M. Leuze, M. Hanack, *Eur. J. Inorg. Chem.* **2001**, 679.

[158] M. Hanack, J. Osio-Barcina, E. Witke, J. Pohmer, *Synthesis* **1992**, 211.

[159] D.D. Perrin, W.L.F. Armarego, Purification of Laboratory Chemicals, **1980**, Pertgamon Press, 3rd Edition, Oxford, New York, Beijing, Frankfurt, Sao Paulo, Sydney, Tokyo, Toronto.

[160] T.S. Ertel, H. Bertagnolli, S. Hückmann, U. Kolb, D. Peter, *Appl. Spectrosc.* **1992**, 46, 690.

[161] M. Newville, P. Livins, Y. Yakoby, J.J. Rehr, E.A. Stern, *Phys. Rev. B* **1993**, 47, 14126.

[162] S. J. Gurman, N. Binstead, I. Ross, *J. Phys. C* **1986**, 19, 1845.

I want morebooks!

Buy your books fast and straightforward online - at one of world's fastest growing online book stores! Environmentally sound due to Print-on-Demand technologies.

Buy your books online at
www.morebooks.shop

Kaufen Sie Ihre Bücher schnell und unkompliziert online – auf einer der am schnellsten wachsenden Buchhandelsplattformen weltweit! Dank Print-On-Demand umwelt- und ressourcenschonend produziert.

Bücher schneller online kaufen
www.morebooks.shop

KS OmniScriptum Publishing
Brivibas gatve 197
LV-1039 Riga, Latvia
Telefax: +371 686 204 55

info@omniscriptum.com
www.omniscriptum.com

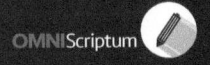

Printed by Books on Demand GmbH, Norderstedt / Germany